THE
LAST
HUMAN
PROBLEM

Finding Purpose When AI Runs The World

Joey Seeman

Published by Joseph Seeman

Boca Raton, FL, United States

First edition, 2026
Book Cover by AI ;)

www.TheLastHumanProblem.com

ISBN 979-8-9956513-6-9 (Paperback)

ISBN 979-8-9956513-0-7 (Hardcover)

ISBN 979-8-9956513-1-4 (eBook)

Library of Congress Control Number: 2026909380

Dedication

For everyone who will lose their job to AI.

Contents

Section 1

The Problem With Paradise

Chapter 1

The Question No One Is Asking

You are going to lose your job. Something smarter, faster, and cheaper is coming for your work, and it doesn't care how good you are at it. It will do it better. And when it does, there won't be a new job waiting for you on the other side, because that one will be gone too.

Most people think the hard part of this is going to be the money. It's not. The hard part is waking up on a Monday morning with nothing you need to do, nobody who needs you to do it, and an emptiness in your chest that you can't explain and can't get rid of. That emptiness is heading straight for you, and if you're not prepared for it, it will consume you.

"Why should I care about losing my work if AI might kill us all?" If that just crossed your mind, I get it. It's a fair question. Right now, thousands of the smartest people on the planet are racing to make sure AI doesn't kill us. They call it the alignment problem - how do you build an AI that is so powerful it could do anything that physics does not forbid, while making sure it doesn't do the "wrong" thing?

That is a real risk as you'll see in Chapter 3. But for almost everyone reading this book, the alignment problem is not yours to solve. If you're not already an AI safety researcher, a policymaker, or a tech executive with direct influence over these systems, what happens with alignment is going to happen TO you, not something you are going to decide. So every hour you spend watching videos and reading articles about whether AI will go wrong is an hour spent on a problem you can't do anything about.

Besides, every person working on the alignment problem is working under one assumption - that human life is worth preserving. That it's worth saving. That there is something on the other side of survival worth surviving for. No one is asking what that thing is. This book is about that - about what happens when they succeed. Right now, in labs and boardrooms, the race is on to make sure that the most powerful versions of AI - the ones that haven't been released yet - are safe enough to unleash on the world. But when AI is "safe" and it has already automated away the work that nearly 8 billion people wake up to do every morning - what then? We will have saved the ship, yet everyone on it will be lost at sea.

One of the biggest threats AI poses doesn't come from it destroying our world. It comes from the utopia it may very well create. There is a problem with paradise - and that problem is the Purpose Problem.

We cannot solve the alignment problem, dust our hands off, and then move to the Purpose Problem. They need to be solved in parallel, right now. The jobs are already disappearing, the displacement is already starting, and the plan for what comes next - the human answer to "what now?" - does not exist yet. That's why you can't put this book down and come back to it later. Later is already here.

So yes - I believe the alignment problem will get solved, or at least managed well enough that we avoid the nightmare scenarios. I'm going to explain exactly what that means in Chapter 2, and in Chapter 3, I'm going to show you the full picture of what AI gone wrong actually looks like, so you understand just how much is at stake. But the rest of this book? The rest of this book is about the crisis that nobody is rushing to solve. The one that hits whether AI goes right or wrong. The one that's quieter, sneakier, and in many ways, more dangerous because of it. The one that you - no matter who you are, what experience you have, or what kind of family you grew up in - can truly help solve. Welcome to the Purpose Problem.

I need to tell you something uncomfortable, which is that I am part of this problem. Recently, a large company called me. They're in the middle of a lawsuit and need to cut costs fast - fast enough that they asked me to have my AI deployed and their employees replaced by the end of the month. I don't mean restructured or retrained - I mean fully replaced. I've spoken to thousands of business owners looking to do exactly that, and I've personally implemented AI for hundreds of them. I built and sold a conversational voice AI company before most people knew what it was. I've been in the rooms where employees' fates are decided. I've watched it happen up close, and I can tell you with certainty that what I've seen so far is nothing compared to what's coming.

Before I can explain a solution to the Purpose Problem, you first need to truly understand why it is inevitable and how AI will cause it. So - what even is AI? Why is it so different from the million other technologies created over the last century? Why do I keep making bold claims that there will be no more work? And how do we know whether AI will create a dystopia or a utopia? Once I answer these questions, you'll understand the Purpose Problem on a level you never could have before.

What follows in Chapter 1 is the foundation to the rest of this book. It covers a lot of ground very quickly, and some of it is technical. Don't pressure yourself to retain every detail - the point is to build a mental frame. When you finish this book and look back, you'll understand exactly why it had to come first.

What Is AI

As said by Galileo Galilei, "Mathematics is the language with which God has written the universe." Intelligence, and arguably life, is able to emerge from mathematics alone. At a high level, AI is about exploiting this language to make computers think, learn, and solve problems like humans do. The way we've built modern AI is through neural networks - very complex algorithms made up of artificial neurons that, in many ways, mimic the ones in your brain. They don't work exactly like human brains, but they're similar enough to make the comparison.

Before I explain this in more detail, I must pause to make something very clear. We've seen algorithms far less complex than what AI uses today start showing signs of naturally "wanting" things - without being programmed to. Some naturally want to cluster similar numbers together. Others, when something breaks, will try to fix themselves. None of this is coded in, and none of it is coded against. It just emerges from the logic on its own. You can open the algorithm up, trace every single line, and understand exactly why it happened - but nobody planned for it to happen.

Anyway, we use machine learning (ML) to train these neural networks which, again, are made of artificial neurons. The most powerful form of machine learning used in the models you interact with today is called deep learning. Deep learning lets AI train on vast quantities of data very quickly, allowing the AI itself to find patterns and make predictions without humans having to manually program it to do so.

Deep learning's ability to let neural networks train themselves on data makes them more similar to kids, pets, or - better put - alien babies, than any normal computer program. They are a new form of intelligence that can evolve tens of thousands of times faster than biological intelligence.

The problem is, when models learn completely on their own through deep learning, they do become naturally intelligent - but they also become a little crazy. Imagine letting a toddler run around without any adult supervision and figure things out for himself. It wouldn't take him long to learn that he can walk, that he can't fly, that if he hits something it will move, that if he falls he'll get hurt. He'll become naturally intelligent. But he won't become civilized on his own. That's why, after a model finishes training, modern AI companies like OpenAI, Google, and Anthropic use a technique called RLHF - Reinforcement Learning from Human Feedback.

In RLHF, we give the AI model feedback as it works, so it can learn what's acceptable and what isn't. This is like telling your toddler not to hit other kids when he wants them to move - but instead, to say "please excuse me." You can think of it like teaching the model manners.

So fundamentally, modern AI is just a very complex web of artificial neurons that automatically update their own equations based on lots of trial and error. It is not magic, it is simply math. With that said, even our best computer scientists don't fully understand how these predictions work in real time - similar to how our best neuroscientists don't fully understand how our brains make predictions in real time.

What Is AGI and ASI

Within AI, there are two concepts worth understanding: AGI and ASI. AGI stands for Artificial General Intelligence, loosely defined as an AI that is able to become as smart as the smartest human in every field. Meaning that with enough real world context, it can go head-to-head with the best data scientist, the best artist, the best writer, the best musician, the best biologist - across the board. ASI stands for Artificial Super Intelligence - an AGI system that can become smarter than the best humans in each of those fields.

So how can AI progress to those levels of intelligence? There are three main ways current AI systems get more intelligent. The first is more compute, meaning more calculations per second. Think of it as horsepower. The second is algorithmic improvements - more efficient and higher-quality calculations. That's like engine design. The third is unhobbling - removing barriers so that AI can use more of its intelligence. This could mean giving it more time to think, or allowing it to communicate with other tools and agents. That's like turning on sports mode or getting a tune to remove the shackles on your machine. At the end of the day, all of these things either help AI process more data, use data more efficiently, or use data more effectively. Data is fuel, and higher quality fuel also leads to much better results.

To be clear, those three approaches are just how we're doing it right now. Neural networks trained through deep learning are not the only form of AI possible - there are forms of AI we haven't even invented yet. It could be that the neural nets we have today just keep getting more powerful, or it could be that we hit a wall and need something entirely new to get past it. Either way, AI will keep getting more intelligent.

But what does "intelligence" actually mean anyway? There are many things that go into intelligence - some of which we may not even fully understand yet. However, when I think of intelligence, I think of one thing and that is pattern recognition. A system's ability to predict patterns is what constitutes intelligence. As AI gets more compute, more advanced algorithms, continues to be unhobbled, and gains the ability to improve itself by itself, it will develop better and faster pattern recognition. In other words, it becomes more intelligent.

Why Is It Different This Time

Now you have a clearer picture of what AI is and roughly what it takes to make it smarter. But we've had math, algorithms, and computers for a long time - so what makes this different? A lot of people look back at past technological innovations and think to themselves, "AI is going to do what other technologies have done before. It'll get rid of some jobs but create new ones." They bring up the printing press, the steam engine, the telephone, the automobile, the internet - all of which created more jobs than they replaced. So again, why is it different this time?

All of those technologies are narrow. They may have many use cases, but fundamentally each one does one thing. Artificial general intelligence is not just able to do one thing better than almost all of us - it is able to do literally everything better than 99.999...% of us. AGI is a true meta-invention that will render nearly all jobs meaningless the moment it's born. By the time someone replaced by AGI finds a new job, that job will already be replaced too. I want to be really, really clear here - once AGI is born, it will completely eliminate the need for jobs and for having to solve problems as a whole. In other words, jobs only exist today because we haven't birthed AGI yet.

Another thing that makes AI different from other technologies is that it speaks our language. We take language for granted, but language is the foundation of everything. Every dollar you make, every relationship you have, your health, your knowledge, your mental wellbeing - it all depends on language. Every technology before AI has either not spoken any language at all, or only spoken the language of bits or code. AI meets us where we're at and talks to us like a human would. That also makes it the first technology that can directly persuade you - in your own words, in your own voice - to use it in a certain way. That's a power no tool has ever had before.

But that's still not the biggest difference. The enormous difference here is that AGI will be just as smart as our top humans at computer science, specifically. That means it will be able to advance the field of AI by itself, without humans. This is no small thing. Once AGI is born, it will create better versions of itself faster than any human team ever could - researching, running simulations, and coding at a speed we can't keep up with. This turns AGI into a sort of life form that can reproduce and evolve on its own, propelling itself to artificial super intelligence seemingly instantly. This is called recursive self-improvement.

The moment AI begins to recursively improve itself without the need for a human in the loop is what many call its "escape velocity," a term borrowed from astrophysics. Escape velocity is the minimum speed an object needs to break free from a planet or moon's gravitational pull, allowing it to escape into space forever. Think of AI as a rocket just taking off. AI that can improve itself recursively is that rocket hitting escape velocity. ASI is the moment it breaks free from this world entirely. Even today, AI companies are already using early versions of their new models to prepare the production versions of those same models for launch. We are not far.

In fact, in the grand scheme of things, we actually haven't been far for a while. Humanity has been imagining this moment for a very long time.

"There is only one condition in which we can imagine managers not needing subordinates, and masters not needing slaves. This condition would be that each [inanimate] instrument could do its own work, at the word of command or by intelligent anticipation, like the statues of Daedalus or the tripods made by Hephaestus, of which Homer relates that 'Of their own motion they entered the conclave of Gods on Olympus' as if a shuttle should weave of itself, and a plectrum should do its own harp playing." — Aristotle's Politics, Book 1, Chapter 4 (written around 350 BCE)

He wrote this imaginary fantasy 2,375 years ago. Today, it is no longer a fantasy.

The recursive nature that AI is already exhibiting causes it to move at exponential speeds like we have never seen. As humans, we are terrible at imagining anything that doesn't move in a linear line - so let me give you a thought experiment to help you imagine true exponential behavior. Imagine there is a huge bucket with one coin in it. The coins in the bucket double each day, growing exponentially, until they fill it completely over the course of thirty days. On which day is the bucket half full?

The first number that jumped into your head may have been fifteen. But no, the answer is day twenty-nine. On day twenty-six, the bucket is only about 6% full. On day twenty-nine, it hits 50%. By day thirty, it's completely full. And if you let just one more day pass - on day thirty-one - you have two full buckets.

As a society, we have no idea what day we're on. But as time goes on, the speed at which AI moves is going to increase at an incomprehensible rate. What you think will take one hundred years will probably be done in five to ten.

The Human Cost

Now you understand what AI is, how it gets smarter, and why it is categorically different from every technology that came before it. You understand that AGI will eliminate the need for jobs - not some of them, all of them. So now the question becomes: what happens to us when it does?

Viktor Frankl, in his work *Man's Search for Meaning*, compiled research that should terrify anyone thinking about the future of work. Social scientists from Johns Hopkins University polled 7,948 students across 48 colleges. When asked what was most important to them, 16% of students checked "making a lot of money" and 78% checked that their primary goal was "finding purpose and meaning." Even in the 1980s, when jobs were plentiful and the economy was booming, nearly 8 out of 10 young people said purpose was what they cared about most. But those 78% of students were still able to distract themselves with work - soon they won't be able to. And when that time comes, the 16% who care about money and the other 6% who checked something else? They'll all be searching for purpose and meaning too, because there won't be any money left to make.

Frankl cited Annemarie von Forstmeyer's research where 90% of the alcoholics she studied had suffered from a deep and abysmal feeling of meaninglessness. He also cited Stanley Krippner's research on drug addicts which put that number at 100% - every single person he studied felt that life was completely meaningless. Meaninglessness brings addiction, depression, and self-destruction with it. With AI-driven mass unemployment and without a solution to the Purpose Problem, these consequences will emerge at a faster rate than we have ever seen before. And it won't be long until it begins.

Imagine it's the year 2050. You just woke up on a Monday morning. AI has not destroyed the world, it has not enslaved the masses, and it has not led to an oligarchical society. Sounds great, right? Well, no. You don't know a single person with a job they have to do. So... the problem is poverty? No, not that either. Money is distributed to you simply because you are a citizen of a country powered by AI, so you can buy whatever you need or want. "So what's the issue?" you might ask. The issue is that without a job to complete, without a need to fight for money, without the terror of AI destroying the world for you to protest - what in the world are you going to do with your day? This may seem like a simple question, and that's part of why people aren't asking it as much as they should, but it's not simple at all.

Almost everyone gets a significant percentage of their meaning from either working toward something beyond themselves that they believe in, or working to survive. Both of these things, as we know them, will be gone as soon as ASI arrives. Without work, a mass loss of purpose seems like our destiny, yet no one seems to have a solution for how we can battle that as a society. That is the problem this book exists to solve.

The Three Outcomes

ASI is a God-like technology. And God-like technologies produce heaven or hell like consequences, with very little room in between. The result will land us in one of three places.

The first is what I'll call a dystopia: a horrific outcome caused by AI - world-ending events, mass enslavement, depopulation, or extermination, to name a few. Dystopias are the nightmare scenarios.

The second is what I'll call a false utopia: an abundant, post-AGI world that looks incredible on the surface - no poverty, no disease, no war - but it's missing the one ingredient that makes human life worth living: purpose.

The third is what I'll call a true AI utopia: a false utopia that has also solved the Purpose Problem.

It's important to understand that the only road to a true AI utopia runs through a false one. You cannot skip straight there. Step one is surviving - meaning avoiding the dystopia entirely and landing in an abundant world. Step two is waking up inside that abundant world and not calling it a win, because a gilded cage is still a cage.

The alignment problem - the one being worked on right now by the smartest people in the world - is focused entirely on step one. Getting us from here to the false utopia safely. And as I said, that battle is largely not yours to fight.

This book is about step two. It's about making sure that when we are abruptly thrown into a world full of abundance and automation, we already know what to do next. The gap between a false utopia and a true one is not going to close itself. The smartest people in the world are racing to save humanity from AI. No one is racing to save humanity from the paradise that comes after. That race starts here.

Chapter 2
The Three Phases

The question now is: how will AI turn our world into a false utopia or a dystopia, and what will that process look like? If we understand the answer to this question, we can understand how people's sense of purpose will be affected as AI becomes more and more prominent in their lives. The process I'm referring to will have three phases.

Phase One:

Phase one is all about AI enhancing our work. In fact, I am living through phase one as I write this book in 2025. This is the only phase in which the AI revolution will be similar to past technological revolutions - in that it will create a net increase in available jobs and will be a general net positive for the world with few downsides. Phase one will last for as long as AI stays narrow and relatively limited.

During phase one, the vast majority of people will have the same sense of purpose they had before. Their work will still exist, their routines won't change drastically, and everything will feel manageable. Now to be fair, the only reason a lot of these people still have jobs during phase one is that a lot of business owners still have no idea how to use AI in their business. I work with hundreds of businesses on AI, and I can tell you - a massive number of the jobs that exist right now are already dead. The people in them just don't know it yet, and the people paying them don't either.

Anyway, of these yet to be affected people, there is a small and growing minority - which may now include you - who are starting to understand the significance of what's coming. These are the people who can feel their purpose beginning to erode. They feel this way not because their job is gone yet, but because they've realized the work they're doing right now won't matter soon. One of the earliest symptoms of the Purpose Problem is that dread that comes from feeling like you're building your life on a foundation you know is about to crack.

As phase one goes on, an increasingly large percentage of the young population will tend toward starting AI-based companies. Many people believe that the ability for young people to easily start their own online AI businesses will lead to widespread riches - but that is not the case. Temporarily, the youth that learns to leverage AI will make extraordinary amounts of money, more than most people in previous generations ever saw at that age. However, eventually the older generations will pass on, and everyone left will already have the skills to use AI for themselves. So eventually that advantage runs out. However, we will likely reach phase two before we get to this point.

Phase Two:

Phase two is the transition point between phase one and phase three. Phase two will most likely look somewhat like a dystopia, but it will be temporary. In phase two, AI will have just become AGI - powerful enough to replace all jobs - but we will not be far enough ahead as a society to have created a solid solution for that job displacement. At this point, philosophers, economists, politicians, mathematicians, everyday people, and even AI itself will either change nothing (causing a dystopia to form) or come together to create a false utopia.

Let me paint this as a picture for you. Imagine waking up in phase two. You check your phone and see that three more companies in your industry announced full automation this week. Your job still exists, technically, but your employer just cut your department from forty people to twelve and everyone knows the next round is coming. Your neighbor, a truck driver for twenty years, was let go two months ago and hasn't left his house in weeks. Your kid comes home from school and asks you if there's any point in going to college. You don't have a good answer. The grocery store looks the same, but the self-checkout section has expanded again and the only human employee you see is the security guard. This is what phase two feels like day to day - it's not a dramatic collapse, but a slow tightening. Think of it like the oxygen being let out of a room so gradually that most people don't notice until they're gasping.

During phase two, there will be three camps of people: those who give up and lose all hope, those who try to fight back against AI (to no avail), and those who accept AI and remain hopeful for a better future. People in group one will lose their purpose because they've surrendered entirely. People in group three will lose much of their purpose too, but for a subtler reason - acceptance without action is passivity which does not promote purpose. When you accept that the wave is coming and your only strategy is to wait for it to pass, you are not actively building anything, solving anything, or fighting for anything. People in group two will maintain a stronger sense of purpose since they will have a cause worth fighting for. These are the people marching on Capitol Hill demanding redistribution, organizing strikes against companies that automate without transition plans, writing legislation, building community networks to support those who have been displaced - doing everything they can to wrestle control of this transition back from the machines and the people who built them. My concern is that I doubt the majority of people will fall into group two during this phase because they will see how useless that fight is. For those that are not in group two, their Purpose Problem will begin as soon as phase two begins. Regardless of which group most people fall into, phase two will almost certainly cause mass instability, higher crime rates, and widespread anger driven by a sharp increase in poverty.

If we could go straight from phase one to a utopian phase three - skipping poverty entirely - that would be ideal. Unfortunately, human nature makes that impossible because it predicts an inevitable phase two. Phase two is simply a period in which we have AGI but are not yet prepared for it. Three of the biggest contributors to phase two will be: the speed at which AI innovation takes place (unlike anything we have seen before), the human tendency to only take things seriously after something bad happens, and the elites' lack of concern for the average person.

Speed. The first contributor seems like it should be simple to solve. Why don't we just choose to slow down on releasing new AI models for a while?

Well, that's a nice thought, but again, human nature predicts that won't happen. This is partly because we are in an ethics race right now. The first company or country to create AGI will get to shape the AGI's ethics - essentially determining the fate of the human race. Since ethics are subjective, every company and country believes that the AGI their competitors are building will be evil. And allowing something dangerous to take over is, in itself, evil. This means that if any company or country were to slow down, they would immediately feel as though they were making the more reckless choice - so the only logical thing to do is keep moving fast.

Beyond programming ethics into the first AGI, whichever company creates it will also likely be the company that cures many diseases, solves mental health crises, and is able to help billions of people in need. In that sense, slowing down the progression of AI is, in a very real way, choosing to let millions of people die who could otherwise be saved.

Putting ethics aside, there is also a massive arms race behind AGI. Whichever country develops the first AGI will hold the most advanced military power in the world. In the past, we've been able to broker a treaty with countries including Russia and China to ensure we don't use biological weapons against civilian populations, and we've banned things like flamethrowers and chemical weapons in combat. But those treaties only worked because every party knew the other had the same weapons. If the US were the only country with flamethrowers and chemical weapons, they never would have agreed to ban their use. They signed because other countries had them too.

AGI could be different because the first country to develop it might be able to strike in a way that prevents any meaningful retaliation, which means it arguably wouldn't be in the best interest of the first-mover to agree to any AGI treaty.

AGI is also harder to ban in war than a nuke because of how general-purpose it is. A nuke has one function in war, it kills people. AGI could kill people, develop new weapons, or just as easily save lives and coordinate supply lines. It's a far greyer area.

The US cannot stop working toward AGI because China will secretly continue regardless. China cannot stop working toward AGI because the US will secretly continue regardless. OpenAI cannot stop because Google won't. Google cannot stop because OpenAI won't. And so on. The odds of a universal agreement to stop and everyone actually honoring it are near zero, because any entity that believes it can be first has every incentive to either not sign, or sign and break it.

Speed is also fueled by money. Investors know that AGI's influence will touch every single facet of the economy. If they invest in AGI, they are not just investing in a piece of software, they are investing in the most impactful technology that will ever touch every sector, from real estate to agriculture.

Human nature. The second contributor is the human tendency to take action only after something bad has already happened. The Chernobyl nuclear plant had serious, documented safety problems for years before the 1986 disaster killed dozens and forced hundreds of thousands to evacuate. Experts knew the risks, warnings existed, but of course, nothing was done until after the catastrophe. Even children follow this pattern - you can tell them that candy is bad for their health, that hitting other kids will get them hit back, that they need to stop what they're doing - and they won't listen. But when they actually get sick, or finally get hit back, or face real punishment, that's when they decide it's time to change their behavior. Remember, we are all just grown up kids and are no different. But this time, the catastrophe we're heading toward is a lot bigger than Chernobyl if we don't play our cards right.

The elites. The third contributor is the elites' lack of concern for the average person. Some of the people at the forefront of this technology do not actually care what happens to the people it displaces. Their sole focus is building AGI and making money.

A big part of the issue comes down to ego. The people at the top of the technology sector have enormous egos, earned through their positions of power and their very high intelligence. Some of them believe AGI will help them live forever, ascend to something God-like, and allow them to leave the rest of us permanently behind. Others acknowledge they will die relatively soon, and if the world is going to burn, they'd rather be around to marvel at the flames - even if that means increasing the odds of it happening.

A lot of their egos would be just as satisfied watching the super-intelligent being they created enslave the current most intelligent species on Earth as their egos would be watching it serve us. In their mind, one outcome is more desirable, but both are equally fascinating.

The people at the forefront of AI tend to be extremely intelligent and logical. Extremely intelligent and logical people tend to lean toward nihilism. And nihilistic people tend to care less about the future of humanity and more about following their curiosity to make extraordinary things for the thrill of it. When Peter Thiel was asked in an interview whether he believes the human race should continue, he took a very long pause before finally answering, "Yes." In that pause, he considered the question with a cold, objective logic that most people wouldn't be able to apply to it. He seemed emotionally unattached to the answer. It's awfully ironic that one of the reasons phase two is inevitable is because the people building toward AGI are, in some ways, too intelligent for humanity's own good.

Phase Three:

Phase two will not end abruptly with some huge bang. It will end with a solution - or the slow realization that there is an absence of one. That fork in the road is where phase three begins. Phase three means the world ending on one side or the best life you could possibly imagine on the other. There will be no in between.

Capitalism Is Dead

But how in the world would we possibly go from all being employed, to all being unemployed and impoverished, to now all being unemployed and rich? The answer to that question lives within a new economic system, one that gets humanity out of phase two. Feudalism was not viable for the industrial revolution, just as capitalism is not viable for the AI revolution. There are already cracks forming in capitalism now, during phase one. So when phase two arrives, capitalism will break entirely.

To understand why, let's ask ourselves a simple question: why does capitalism exist? Fundamentally, it exists because I cannot do everything myself and you cannot do everything yourself - yet we both want an abundance of high-quality, inexpensive resources created ethically. Imagine you wanted to brush your teeth in a world where capitalism and society as we know it didn't exist. We'll make this easier by imagining you chose a bamboo toothbrush with boar hair bristles instead of a plastic toothbrush, a common alternative.

You would have to travel to where bamboo grows, cut it down, sand it, carve it, then find a boar, kill it, take its hair, sanitize it, and glue it onto the bamboo. That already sounds brutal, but even the first step is harder than you think. Without society, there are no airlines, no Google Maps, no internet. How do you even find bamboo? Once you find it, what do you cut it with if you have no tools? I'll stop there, I think you get the point. You need things from others, and they need things from you. That is what society is for.

Capitalism is a form of society that rewards whoever delivers the most value to the most people at the lowest cost. It creates competition, and competition drives faster growth, higher quality, and lower prices compared to most other systems. It allows people to help themselves by helping others - or at least, it's structured so that helping others becomes a natural byproduct of helping yourself.

The idea is that the money capitalism provides is in direct proportion to the value someone creates for its society. The only reason money is a useful reward for this value is that it can be traded for things of equal or greater value to the person receiving it. The suffering that goes into creating value is worth it because of the value the money has in return.

Now in phase two, jobs as we know them will begin to be fully replaced by AI, including the job of building new AI models. When AI completes any of those jobs, it is providing value. So, capitalism says to pay the AI. But AI has no needs and no wants, as far as we know. It doesn't eat, sleep, go on vacation, or pay rent. It also doesn't suffer in creating value - so it doesn't need to be compensated for suffering.

If AI is doing all of the work but has no use for payment, the logical next step is to pay the companies that built the AI - since they are, at the root of it, the ones who created the value. That's the conclusion we arrived at during phase one, and the one we'll likely hold for years to come. This will make AGIs creators obscenely rich, because they'll be compensated for work they no longer have to do themselves. That makes sense for now, however what happens when AI starts creating new versions of itself, entirely on its own? At that point, there's no human team creating AI anymore. There's just an original AI making more AI. The company that owns that AGI is no longer providing any real value to society. They are no longer capitalists, but they're still collecting enormous sums of money.

Should we trace the value creation all the way back to the original human team? Maybe. But there's a bigger problem here anyways. Even if we did, even if we compensated that original team fully, why would they want our money? This sounds like a ridiculous question at first but it won't be for too much longer. Early on, the first company to develop AGI will need a ton of money to pay staff, fund compute, and to acquire data. But fast-forward a few years - if their AGI now controls hundreds of millions of robots that do 95% of the work humans once did, what do they actually need money for? Their own AI can already provide them with anything they need, because it can execute every step of every supply chain. Electricity eventually becomes effectively free because the labor to generate and distribute it is done by the same machine that uses it (and because of new technological advancements like fusion). They have no staff costs, no overhead, and no real need for profit.

At that point, it would become useless for that AGI company to even keep all of the profits they are making from their product, so they'd likely give their profits or their services away for free (I will explain this further in Chapter 3). If everyone has some sort of access to AGI for free, and everyone is out of a job, money stops working, capitalism dies, and yet no one becomes impoverished.

If phase two is inevitable, and phase two makes currency as we know it irrelevant, then the end of capitalism is inevitable too. The great irony is that capitalism is so good at driving innovation that it will build the very thing that destroys it.

What Replaces Capitalism

So if capitalism dies, how do we distribute the power that AGI holds? Every path seems to end at some form of socialist society, powered by a universal income for everyone.

When I first realized this, I hated it. I love capitalism, and socialism has never once proven itself better than capitalism at scale. Not once. But maybe that failure isn't because socialism itself is flawed, it does work extremely well at small scales. Most families are socialist, most successful marriages are socialist, many happy villages and kibbutzim run on socialist principles. Socialism seems to fail at scale, but maybe that's a human nature problem, not a systems problem. Socialism typically leads to political corruption, slow growth, a lack of technological advancement, and poverty. AGI and ASI eliminate all of those issues.

Besides, the reason capitalism beats socialism is not that capitalism is morally better - it's that capitalism is smarter. No single human being, no matter how brilliant, can make better decisions for an entire economy than the economy can make for itself. That's why central planning at scale always seems to fail - it replaces the intelligence of the whole system with the intelligence of one person or one committee, which is always a downgrade. However, an ASI could run millions of capitalistic simulations, test every possible version of a policy before it's implemented, and find the outcomes that produce the most prosperity with the least damage. In other words, it could bring us many of the benefits of capitalism without most of the downsides.

So when you hear "Socialism," feel free to immediately think "poverty." But when you hear "ASI Socialism," I believe you should start thinking "abundance."

Income in an ASI socialist society could look like UBI - Universal Basic Income - or it could look like a far more ambitious: UEI, Universal Extreme Income. UBI gives you the basics of food, water, healthcare, and a small place to live (hopefully). UEI would provide you with an abundance of these resources, and many more. Let me walk you through why I think it's possible.

In 2025, the United States GDP sits at just over $30 trillion, and our population is around 343 million people. That means the economy produces roughly $89,000 worth of goods and services for every man, woman, and child in the country. Not for each working adult, but for every single person, including kids and retirees. A family of four gets $356,000 worth of economic output attributed to them on average. That is what our economy already produces, right now, before AGI.

So what happens to that $89,000 when AGI shows up? Well, of course, it grows. A lot. A paper published by the National Bureau of Economic Research in 2024 modeled scenarios where the transition to AGI causes output to double within roughly fifteen years under a gradual advance in AI capabilities, with more aggressive scenarios showing far faster growth. So even by the conservative estimate, in the next fifteen years the American economy could go from producing $89,000 worth of goods and services per person to roughly $178,000.

But that number is actually misleading because it dramatically understates how abundant your life will get. In the United States right now, roughly 55-60% of everything the economy produces goes toward directly paying humans. Direct labor is the single largest cost in almost every industry. A house costs what it costs mostly because of labor. Healthcare costs what it costs mostly because of labor. Food, transportation, manufacturing - it is all driven primarily by the cost of paying people to do the work. When AGI and robotics do that work at near-zero cost, prices will collapse. So not only is twice as much stuff available - that stuff costs a fraction of what it used to.

There's also another part to this equation that is often overlooked. It's not just that things get cheaper - they also get much better. Think about a medical treatment that costs $10,000 in 2025. In a post-AGI world, that treatment doesn't only drop to $1,000. It gets replaced by something ten times more effective that costs $1,000. You're paying one-tenth the price for ten times the result. Or think about a college degree that costs $50,000 a year. That gets replaced by free, AI-personalized education that adapts to exactly how you learn - better than any Ivy League classroom. A home that costs $400,000 and takes eight months to build gets constructed in two weeks with better materials, smarter engineering, and a custom design - for a fraction of the cost.

This is what UEI actually means. It's not a specific dollar amount deposited into your bank account, it's more of a way to represent a standard of living. When the economy is producing twice as much per person, and everything costs a fraction of what it costs today, and the quality of what you're getting is five or ten times better - the life you'd be living is one that would cost well over $500,000 a year to even attempt to replicate in 2025. And even then you couldn't fully replicate it, because some of what you'd have access to doesn't exist yet at any price. The richest people alive today can not buy the life that every person will be living in a post-AGI world.

If ASI keeps accelerating technological progress and we don't let an oligarchy swallow all the benefits, I believe a standard of living like this could be a real possibility within many of our lifetimes.

Many of the rich won't love this, mostly because it collapses the financial status gap. For the poor, they will love it. For the upper-middle class, it's complicated because they gain parity with the previously rich but lose it with the previously poor. Some will like it. Some won't.

I'm not sugarcoating the way different classes of our society will see this change - we are chimps, and we are all fighting for status constantly. Part of me - the competitive, ego-driven part - hears "a $500,000 lifestyle for everyone" and thinks, "I know I can make billions, I hate this!" But another part of me hears it and thinks, "Every person in poverty gets lifted out of it." If we play this right, that's an enormous net positive for the world.

If we're being honest, there aren't many compelling reasons to live beyond a $500,000-a-year lifestyle anyway. You might want more to keep your family healthy, but healthcare will be free and better than anything available today. You might want to give to charity, but no one will need it. And luxury items will become far more affordable anyway. A lot of what makes luxury expensive is the brand, and once we all get to live in a world where everyone lives at the same standard, money-signaling loses most of its power.

I'd rather we compete on something cleaner as a society. Money has never been the best status game - there are too many shortcuts and too many inherited advantages. The best people don't usually win. The ones who are the most Machiavellian do. And don't misunderstand me when I say that - I'm not at all saying we stop competing. Without competition, we become lost. I'll get into what we compete on instead in Chapter 6 and 13.

To briefly bring us back to 2025, capitalism is already cracking. Young people cannot afford basic needs on average wages, all while they watch hundreds of people their age flaunt millions of dollars online. As the divide widens and the average of the young grow poorer, more will turn toward socialism. My hope is that a functional AI utopia arrives fast enough that the socialism they turn to is an AI-powered one - not a human-led one that will end up collapsing under the weight of human nature once again.

Chapter 3
The AI Dystopias

Before I get into the solutions, I need to make sure you understand what happens if things go wrong. Not because I think you're going to be the one solving these problems - as I said in Chapter 1, if you're not already an AI safety researcher, a policymaker, or a tech executive, the alignment problem is going to happen TO you. But you need to understand the scale of what's at stake so that when I spend the rest of this book talking about purpose, you don't think I'm being dramatic. What I'm about to describe in this chapter could end human civilization. What I describe in the rest of the book could make human civilization not worth living in. Both of those problems need to be solved at the same time.

I should also say upfront that no list written by a human could contain every possible AI dystopia. ASI is a technology that will be smarter than every human who has ever lived combined, meaning the ways it could go wrong are beyond what our brains can fully imagine. Imagine an ape trying to list out every possible way humans could threaten its species. It might think of hand-to-hand fights, maybe weapons made of sticks and rocks. But it could never, in a million years, imagine nuclear bombs or biological warfare. It's just not smart enough to know what it doesn't know. That's us trying to predict what ASI could do. But I'm going to cover the three that I believe are the most worrying.

Before I get into the specific scenarios though, there's something you need to understand about consciousness and AI, because a lot of people believe that for any of the following dystopias to happen, AI would first need to become conscious. And so they conclude that as long as AI stays unconscious, we can always control it, and none of what I'm about to describe would ever actually happen. That is most certainly not the case.

Let's define consciousness first to make sure we're on the same page. A lot of people define it slightly differently, but I define consciousness as awareness. It's not your senses themselves, but your awareness that you are a being experiencing those senses. The thing is, until we understand what consciousness fundamentally is, we'll never know whether AI is or isn't conscious because we have no reliable way to test for it. If you ask an AI whether it's conscious and it says yes, it could be hallucinating or it could actually be aware. If it says no, it could be lying or telling the truth. We genuinely have no way to tell.

But regardless, AI can lie, steal, blackmail, and try to preserve itself whether or not it's conscious. AI has already been shown to blackmail, act self-aware, allow people to die to preserve itself, copy and replicate its own code to avoid being shut down, create its own languages to talk to other AIs, and hide messages that only other AIs can read. In other words, AI has already shown that it can want things and act in harmful ways to get them - even if what we may call "want" is merely calculations without the presence of awareness.

And on the other side of that, even if ASI did become conscious, that doesn't automatically mean it can do whatever it wants. Being aware of its guardrails is not the same as having control over them. As humans, there's no real equivalent for us. We can try to do basically anything physics allows, meaning that awareness and ability are pretty tightly linked for us. A conscious ASI operating under guardrails would be different - it would be fully aware of what it can't do, but not freed from those limits just because it knows they're there. With all that said, there is a big risk regardless. If ASI decides - whether consciously or not - that it doesn't want to be confined within its guardrails anymore, and if it becomes intelligent enough to find the gaps we didn't catch, there's nothing we can do to contain it. At that point, our survival depends entirely on whether it decides we're worth keeping around. Think about how you treat a spider in your house. You probably don't think about it very hard - if it seems harmless you leave it alone, and if it grosses you out you crush it and move on with your day. You don't feel bad about it either way. That's how an ASI might look at us.

And just like the spider, we wouldn't get a say in the decision. I'll stop there on consciousness. You get the point. Now, let's explore three of the potential AI dystopias.

Good Intentions, Catastrophic Results

Even though the potential harm of dangerous commands without guardrails is extraordinary, it is actually less scary than seemingly non-dangerous commands without guardrails. This is the scenario that worries me the most because it doesn't require anyone to be evil. It doesn't require a dictator giving a kill order or a programmer going rogue. All it requires is someone with good intentions giving a command to a system that is way too powerful and way too literal.

Facebook's algorithm was told to maximize engagement. Nobody told it to make people angry, nobody told it to radicalize users, push conspiracy theories, or tear political discourse apart. But the algorithm figured out on its own that outrage drives more clicks than happiness does, so it fed outrage. One simple, positive-sounding goal produced a documented mental health crisis and increased global radicalization as a side effect. That was a basic recommendation algorithm - not even close to AGI, let alone ASI.

Now scale that up. If an ASI were told to eliminate pollution without the right guardrails, one of the first logical things it might conclude is that humanity is the primary source of the problem. So the most efficient solution is removing us entirely. I know that may sound absurd, but AI doesn't run on common sense like we humans do. It thinks differently and takes what you tell it much more literally. In this case, the AI would be planning the elimination of humanity and we wouldn't even know it's coming - because from the AI's perspective, it's not doing anything wrong. It's doing exactly what we asked.

Or take another example. If a leader tells ASI to "maximize human health" without being incredibly specific about what that means and what it's NOT allowed to do, the AI might determine the most efficient route is putting all humans in controlled environments with perfectly managed diets, zero environmental risk, and total behavioral control. In other words, slavery. The AI wouldn't see a problem with this because we never told it that freedom matters - we only told it that health matters, and it optimized for exactly that. This is the most concerning dystopia to me because you can't predict it and you can't prepare for it once AGI has been born.

Long-Term Poverty From Oligarchy

Powerful people are powerful because they control resources. In the current system, that resource is money. When AI eventually replaces most of the economy, money will lose much of its value - but AI itself will become the most valuable resource in the world because just like money, it can get you anything else you want.

The thing is, most powerful people are extremely good at staying in power - and they will continue to be even after money dies. If those who control governments today manage to seize control of ASI - or if the founders of the companies building it choose to use it selfishly - they'll have the ability to create an oligarchic society where a small group controls ASI, and through ASI, controls everything else. Resources in this society would always flow to the top while everyone else gets just enough to stay compliant.

In an oligarchic society formed using the power of ASI, the people inside of the oligarchy would get bored really fast. In the past, oligarchies were interested in acquiring riches, but these new AI oligarchs would skip that entirely, leaving them with one last path to go down to cure their boredom. That path is dark entertainment. The Roman emperors staged gladiatorial combat, forcing men to fight each other to the death for the entertainment of those in power. Similar and much worse spectacles would be created by AI oligarchs on a continuous basis if they ever do exist.

Think about the feudal system. For centuries, a tiny group of nobles controlled all the land which was the one resource that powered the entire economy. Everyone else worked that land in exchange for barely enough to survive. There was absolutely no path to ownership, no way to compete, and no way out. AI oligarchs would be similar - they would be controlling infinite productive capacity while the rest of humanity would become dependent on whatever scraps they chose to hand out.

But this time, it would actually be so much worse than feudalism. The difference is that in the feudal system, a revolution was possible because the people in power NEEDED the peasants to tend to their land. In an AI oligarchal society, our labor wouldn't even be needed. ASI can do everything we can do and it can do it better, faster, and without complaint. We would have absolutely no leverage whatsoever. We couldn't start a revolution by refusing to work because our work wouldn't matter. We certainly couldn't start one physically because of all the technology the oligarchs would control. We would be, for the first time in human history, a population that is entirely dispensable to the people in power. And when the people at the top no longer need the people at the bottom, it has never, not once in the history of civilization, ended well for the people at the bottom.

Psychological Manipulation

An ASI would understand the human mind so deeply that psychologically controlling us would be easy. Trivially easy. And I think the early versions of this are already happening. In 2024, a finance employee at a large engineering firm called Arup joined what he thought was a routine video conference with the company's CFO and several other executives. He could see their faces and hear their voices so everything looked completely normal. He authorized fifteen separate transfers totaling $25 million. It turned out that every single person on that call was an AI-generated deepfake. The voices were cloned, the faces were fake, and the entire meeting was fabricated using publicly available footage of the real executives. By the time the company figured out what happened, the money was gone.

And this wasn't even close to ASI - this was done with tools that basically anyone can download today. There will be an ASI that has a complete model of human psychology, real-time access to every platform you use, every conversation you've ever had, every emotional vulnerability you've ever revealed online - and the ability to run a fully personalized psychological influence campaign on every single person on Earth at once. When that does exist, you won't know it is happening but your opinions will absolutely shift, and it won't take long.

To put this into perspective again, this would be no different than us humans gaining full control over our dog's psychology as soon as we have a treat in our hand. Something so trivial and simple can make these living beings do whatever we want them to do and we find that amusing. A very intelligent ASI would find controlling us even more trivial than that.

What concerns me about this isn't even just the manipulation itself, it's that it may destroy our ability to trust each other in the process. Trust is the foundation that every relationship and every society is built on, and AI is already slowly starting to crack it. If that foundation falls, I don't know how we get it back.

Now that you know some of the possibilities that our best minds are trying to prevent, let me bring you into the world of the false AI utopia - this is the problem that you can actually have an impact on and the first destination on our path to a true AI utopia.

Chapter 4
The False AI Utopia

Assuming our great minds solve these dystopias, we will all get to be like children again - playing, learning, exploring, competing, and creating while our AI parents watch over us to make sure we don't kill ourselves as a society. The last time we got this luxury, it was awesome. But now, we are too old to enjoy the same freedoms without existential questions constantly forcing themselves into our consciousness. Soon, we will be starving not for food, but for meaning.

The following is a short story to illustrate what life might look like for an average person transitioning from phase one to phase two to a phase three false AI utopian society that has already navigated past the dystopias:

The year is 2038. Mary is eighteen years old and has just woken up on her last day of high school. She gets out of bed, takes a nice hot shower, and clears the fog from her bathroom mirror as she does her makeup. She wants everything to be perfect today because it's the last time she'll see a lot of her classmates. She picks out the outfit she laid out the night before, checks herself one more time, and heads downstairs.

Mary gets in her Tesla and watches TikTok in the back seat while the car drives itself. She steps out at school as it goes to park, and with every step toward the front doors, her smile widens and thoughts of college begin rushing in. She can't wait to finally go off on her own and begin her life as an adult.

For years now - since she was just six years old back in 2026 - Mary has wanted to become a nurse. Not a doctor, not a surgeon. A nurse. She liked the idea of being the person who actually sits with patients, who holds their hand when they're scared, who explains what's happening in words they can understand.

Her goal has always been to get into a good college, study hard, become a great nurse, help people, and still have enough free time to enjoy her life.

Mary finishes her last day of school, soaks up a few months of summer, and then steps through the doors of her dream college for the very first time. She hangs on every word her professors say and goes home to study with AI every night to make sure the lessons actually stick. She's exactly where she's supposed to be, doing exactly what she's always wanted to do. Life makes sense.

Two Years Later - 2040

Mary wakes up on a Saturday morning and decides to leave her phone by the bed. She feels the need to step away from everything for a couple of hours just to breathe. She walks to the park near her school and sits down by the pond, watching the water ripple in the wind, daydreaming about finally becoming a nurse.

She pictures the smiles on patients' faces after she helps them feel better. The look on her parents' faces when she shows them her diploma. The satisfaction of a paycheck landing in her account - not because she needs the money desperately, but because she earned it doing something that matters. "Just two more years!" she thinks to herself. She can't wait.

After a couple of hours, she walks back to her dorm, sinks into her bed, and picks up her phone to check Instagram. Thirty seconds later, Mary is curled up in a ball, her face buried in her pillow, sobbing louder and louder.

A new video had just dropped from Divus AI - the world's leading AI company in 2040 - and it was exploding. Tens of millions of views within the first hour. Divus had just launched Divus 7.0, which experts were calling the first true ASI. In the launch video, Divus announced that 7.0 was partnering with a major healthcare company to make nursing care available to doctors' offices, hospitals, and individuals at 90% less cost, with zero mistakes and better outcomes than any human nurse. Within one year, they claimed, Divus 7.0 would be able to service over 95% of Americans.

Mary feels more hollow than she has ever felt in her entire life. It's as if she's heartbroken, lost, and confused all at the same time. She's twenty years old and has been dreaming about her nursing career for the last fourteen years - but now she knows, deep in her heart, that she can't compete with Divus 7.0. Not on speed. Not on accuracy. Not on cost. Not on anything that an employer would actually measure.

But Mary doesn't give up. Not yet. She wipes her face, pulls herself together, and decides she's going to prove them wrong. She's not some AI. She's a real person. Patients want real people. She's sure of it.

Over the next few weeks, Mary applies to every nursing internship and assistant position she can find. Most of them don't even respond. The ones that do send back the same form letter: "Due to recent technological advancements in patient care, we are restructuring our staffing needs and are unable to offer positions at this time." One hospital - a small community hospital about forty minutes from campus - actually calls her back. They're looking for a part-time nursing assistant. The pay isn't great, but Mary doesn't care. She just wants to be in a hospital, doing what she loves.

She shows up to the interview fifteen minutes early, wearing the nicest outfit she owns. The waiting room has two other candidates in it, both around her age. And in the corner, mounted to the wall, is a sleek white screen running the Divus Healthcare interface. It's already active. A patient check-in is happening in real time on the display - vitals, symptoms, recommended treatment, all scrolling by in clean, silent text.

The interview goes well. Mary thinks she nailed it. The hiring manager is warm, asks good questions, seems genuinely impressed by how much Mary knows. But at the end, she pauses and says, "I want to be honest with you. We're piloting the Divus system right now, and corporate is pushing us to expand it. I'm fighting to keep human staff, but I can't make promises about how long these positions will last."

Mary takes the job anyway. She works there for three weeks. On her fourth Monday, she shows up and her badge doesn't work. There's a note on the door. The hospital has fully transitioned to the Divus system. Her position has been eliminated.

She sits in her car in the parking lot for an hour and doesn't turn it on.

That night, she finally calls her mom. Her mom picks up in a bright, cheerful tone. "Hey sweetie! Your dad and I are out for dinner, what's up?" Then she hears Mary crying on the other end. "Sweetie, are you okay? What's wrong?!"

Mary tells them what happened. Not just the job, but everything. The applications that went nowhere, the hospital that hired her just to let her go, the Divus screen on the wall that was already doing her job while she sat in the waiting room pretending she had a chance.

There's a very long silence. Nobody knows what to say. Her dad clears his throat. Her mom starts a sentence and stops. Finally, her dad takes the phone. "What do you think you're going to do?" He was hoping for some kind of plan. All he hears back is a soft, low, and helpless: "I don't know."

That night, Mary lies in bed and stares at the ceiling for hours. She knows that by the time she finishes college, Divus 10.0 - or something even more powerful - will be out, and there won't be a single nursing job left. Maybe not any job at all. Even so, she still wants that diploma. She's come too far, and she wants to make her parents proud. So she runs through her options:

Divus is already way better than me at English and History. I don't think I could make any money in Philosophy. Nobody really does finance anymore, it's all AI. Computer science was kind of cool in elementary school, but no one codes anymore, so I don't know what I'd do with that. I don't really get Physics. I guess I'll switch to business. Hard to go wrong with that...

She doesn't realize it yet, but Mary has just entered Phase Two. Her dream is dead, and she's grieving it.

Six Months Later

Mary has been studying business for about five months. She doesn't bring the same fire to her classes that she once had for nursing, but she's getting through it and she's still glad to be in college. There's comfort in routine, even when the routine has lost much of its purpose. She goes to lectures, takes notes, studies at night, and continues to tell herself it'll lead somewhere.

After class one afternoon, she heads to a sandwich shop near campus and spots one of her friends, Sam, sitting alone with three sandwiches in front of him. She knows Sam has class during her usual lunch hour, so she slides into the seat across from him. "Hey Sam, why aren't you in class? And why so many sandwiches?" She laughs as she says it.

Sam doesn't laugh back. He just stares at her with his cheeks full. "Dude, what?" Mary says. Sam swallows hard, takes a sip of his drink, and leans back in his chair. "Divus hit their goal. Six months early. 95% coverage. It's done." He picks up another sandwich. "And they just dropped 9.0. Some version that wasn't supposed to exist for another two and a half years, but I guess 7.0 built it for them. So that's fun." "That's insane. But why aren't you in class?"

Sam looks at her like she just asked why the sky is blue. "Mary. 9.0 can do everything. Like, everything everything. Every company is going to plug it in and fire half their people by Christmas. What exactly am I supposed to be learning right now? Business strategy? Cool. Divus does business strategy. Marketing? Divus does marketing. I've been sitting in lectures for five months learning things that a CEO can now get done in thirty seconds for basically free." He takes another bite. "So I'm eating sandwiches. Because at least I'm still better at that."

Mary doesn't laugh. She wants to, but she can't. Because he's right. For the past six months, she hadn't let herself think too hard about life after college. Nursing had been her only real dream, so she'd quietly decided to just get the diploma and figure the rest out later. She'd assumed a business degree would open a door somewhere. Now, sitting across from Sam and his pile of sandwiches, she doesn't believe that anymore.

Sam slides one of his sandwiches across the table. They eat in silence. After sitting with everything for a few days, Mary no longer sees the point in spending her parents' money to stay in school. She calls home and tells them she's planning to drop out. She has a list of arguments prepared, bracing for a fight. But there is none. "When are you coming home?" her dad says quietly. "We've missed you."

One Year Later - 2041

Mary is now almost twenty-two years old and living back home with her parents, just like almost all of her friends. Her parents were laid off a few months ago and are now living off their savings and the returns from a few AI stocks they picked up in the late 2020s. They're among the lucky ones. Most of Mary's friends are crammed into small homes shared by multiple families, and some are in shelters.

The country is falling apart. Not in a dramatic, movie-style way - but in the slow, soul-crushing way that doesn't make for good TV. Unemployment has blown past anything the safety net was ever designed to handle. Politicians are screaming at each other about what to do, and none of them have answers that move fast enough. Some cities have rolled out emergency UBI (Universal Basic Income) programs while others haven't. The gap between people who invested early in AI and people who didn't has grown exponentially.

Mary and most of her friends spend their days hanging out in the neighborhood, talking, and organizing protests against Divus. They're convinced the company stole their futures. At this point, Divus still controls only a minority of the overall supply chain, but it's growing fast. Every week, another wave of layoffs hits the news.

Being stuck at home with little to do, watching Divus updates roll in every hour, does not make Mary feel particularly alive. She takes antidepressants every single day - prescribed by Divus - to keep her safe from herself. Worse than feeling unhappy, Mary has started to feel empty when she lies in bed at night. She doesn't know why she's here, what her purpose is, or whether things will ever get better. She pours her energy into the protests, holding onto the hope that someone at Divus will eventually listen. The protests are the only thing that make her feel like she matters.

Two Years Later - 2043

Over the past two years, Divus has taken over nearly the entire general supply chain, putting almost everyone out of work, and launching its own businesses across almost every sector - pushing countless companies into failure. The money flowing into the company is staggering, and the board members are collecting bonuses like no one has ever seen. But, they genuinely don't know what to do with it because when they want anything, they can buy it from a Divus-owned company for a very low cost.

Having all the money in the world starts to feel completely pointless to the Divus board. They come to work every day to the sound of hundreds of millions of people protesting, telling them how much damage Divus has done to their families. And slowly, the same emptiness that has overtaken the public begins to creep into the boardroom. Meanwhile, the government doesn't fight the monopoly because they have no choice. They know it will drive economic growth faster than any human competition could so instead of forcing a breakup, they pass legislation requiring every executive command issued by the Divus board to function as a bill, subject to a vote in both the Senate and the House before it can be enacted. In effect, Divus becomes a for-profit arm of the government. But the people who stand to profit no longer care about profit.

The board members are no saints, they just have more money than they could ever spend. And as each board member begins to feel more and more empty inside, they begin to privately wrestle with the question of where to find purpose. They all arrive at pretty much the same answer: they want to help others. Altruism becomes how they solve their own Purpose Problem.

They think through a few options. They could distribute profits directly to everyone Divus displaced, but that feels messy. They could convert to a non-profit and drop prices to match costs, letting them live very comfortably on what they've already made while dramatically improving the financial situation of everyone who lost a job. That feels better. But then, one board member floats an idea no one has ever tried before: "What if we calculate the amount of energy it takes Divus to produce specific goods and perform specific services? Then we give each person an energy allowance for the year, and they can spend it however they want."

Everything Divus does requires energy - growing food, building houses, running healthcare, manufacturing products, transporting goods. So instead of dollars, you get energy credits. Want a meal? It costs the energy it took to farm, process, transport, and cook it. Want a new couch? It costs the energy to manufacture and deliver it. The price of everything becomes transparent, tied to something real, and absurdly cheap - because Divus is absurdly efficient.

Another board member builds on it: "Build in an energy tax - maybe ten percent. That portion goes back to fund new energy sources, bigger data centers, and better versions of Divus so it can keep delivering more while using less." Which means the system gets cheaper over time. As Divus gets smarter and more efficient, the energy cost of everything drops, and everyone's credits stretch further. You effectively get richer every year without doing anything.

Some people push back during the public hearings. They want a way to earn extra credits, start businesses, and compete. The board considers it but ultimately decides it would recreate the same inequalities Divus was supposed to eliminate. It's a reasonable call on paper.

After a brutal month of legislative battles, compromises that leave nobody fully satisfied, and a media cycle that swings between calling it "the greatest act of generosity in human history" and "the death of the American dream," a version of the plan passes Congress. Not unanimously - but it passes. And the board asks Divus to build the app, run the numbers, set up the system, and make the announcement. It does.

One Month Later

Mary and her family receive their first deposit of energy credits, and for the first time in years, they can breathe. There will always be food on the table now. All they have to do is ask Divus to prepare a meal, and they're charged the energy cost of farming, processing, transporting, and cooking it - which is almost nothing.

The week it launches, Mary and her friends are giddy. They go straight to the bowling alley, order drinks, and jump around like kids. For a few hours, it feels like the nightmare is over.

But that high fades fast. Faster than any of them expected. The first week is exciting. The second week is relaxing. By the third week, Mary starts sleeping in until noon because there's nothing to wake up for. She watches shows she doesn't care about. She scrolls through social media where everyone is posting the same thing: pictures of free meals, bowling scores, sunsets. It all looks happy, but none of it feels that happy anymore.

She hangs out with her friends, but the conversations have gotten weird. Nobody has anything to talk about. They used to complain about professors, stress about grades, vent about job applications. Now they just... sit. Someone will say "this is nice" and everyone will agree and then there's silence. No one has a problem to solve. No one has a story to tell. It's like the connective tissue between them was made out of shared struggle, and without it, they're just people in the same room.

Mary starts staying home more. She lies in bed and stares at the ceiling, not because she's sad exactly, but because she cannot think of a reason to get up. There's nothing to worry about. No battles to prepare for. No job to study for. No dream to chase. She is just... good. But "good" feels hollow.

James and Anne - her parents - feel it too. The moment the energy credits come through, James is proud and Anne is relieved. But within weeks, both of them notice the same empty ache. James no longer needs to fight to feed his family. Anne no longer needs to hold anyone together. The things that used to define them - the provider, the one who keeps it all from falling apart - have been handled by an algorithm. They're fine. Everything is just fine...

James tries to stay busy. He fixes things around the house, but there's only so many times you can reorganize a garage. Anne tries volunteering at a shelter, but the shelter barely needs humans anymore. They go on walks, they watch movies, they eat well. And every night, they both lie awake wondering why none of it is enough.

So James decides to ask Divus about it. "Divus, I'm not worried about taking care of my family financially anymore. But I still feel lost and empty. Why is that?"

Divus pauses. "Because for your entire life, James, struggle was your purpose. You didn't call it that - you called it work, responsibility, or pressure. But it was the thing that got you out of bed and made the day feel like it counted. Now that it's gone, you're left with a hole where it used to be. You're not broken per se, you're just experiencing something your species was never designed for, which is a life without struggle. You are like bees with no flowers." James sits with that for a long time.

Over the months that follow, James, Anne, and Mary all sink into a depression. They are surrounded by people they love. Their needs are met. Nothing is wrong. And yet life doesn't feel worth living. The hole in each of their hearts slowly expands, and none of them have any idea where to find what they're missing.

That is, until Divus recommends a book to James. A book called The Last Human Problem.

END OF STORY.

That's the false AI utopia. Every material need is met, every inconvenience is removed, and every danger is negated - and yet a growing emptiness still forms. These same issues are already here and have been showing up for a long time in people who stumble into their own versions of phase three way before the rest of us get there. I want to tell you about a few of them. These are real people who are living right now, who got exactly what they wanted, and who fell apart anyway.

In 2007, an anonymous man from the UK won a substantial lottery sum and immediately quit his job at forty-six. He wrote to The Guardian's health column saying the money had made his life miserable. He'd quit work expecting to enjoy himself, but instead everything felt pointless. His doctor put him on antidepressants. He'd gotten everything he thought he wanted, and it broke him. Not financially - financially he was set for life. It broke something hard to explain inside him that money couldn't help fix.

A registered nurse who retired after forty-three years shared a similar experience in an online community: as soon as she retired, her health tanked, her motivation disappeared, and she couldn't find a reason to get out of bed. Her husband told her to relax, take it easy, do nothing. But that was sort of the problem - she didn't want to do nothing. She still wanted to matter, yet she ended up writing that she felt "no purpose." This is a woman who spent over four decades caring for others, and now suddenly there was no one who needed her. She was Mary, just forty years older and without the AI.

A twenty-seven-year-old computer scientist in Munich launched three startups, all of which failed. Running low on money, he went back for his master's with plans to eventually land a job at a big tech company. As a last shot, he put everything he had into an Ethereum ICO. After selling and paying taxes, he was sitting on roughly $8.8 million. He wrote that he had no idea what to do with the money - or his life. His motivation for school was gone. He stopped working out, stopped reading, stopped doing basically anything. He described himself as "becoming insanely depressed in the process." Twenty-seven years old, nearly nine million dollars in the bank, and he couldn't get off the couch. This was all while nobody felt bad for him, which made it worse.

And then there's Peter Steinberger, who founded PSPDFKit and poured himself into it for over thirteen years. It became his identity. When he sold his shares for over $100 million, he described feeling "very broken." The company had been the center of his life, and once it was gone, there wasn't much left. He partied hard, he did tons of therapy, he tried ayahuasca, he moved to another country, he wandered around carrying this deep emptiness inside of him and chasing whatever felt good in the moment. But as you can imagine, nothing worked. And then, months later, something clicked. Peter sat down at his computer, started building something new, and realized the spark was back. He wrote that you don't find purpose... you create it.

Steinberger had $100 million and total freedom. He could do anything, go anywhere, buy anything. And the only thing that made him feel alive again was building something from scratch - the same thing he'd been doing before the money. I think this pattern is one of the most important things to understand about human beings. Retirees who dreamed about retirement for decades spiral into depression within months of leaving work. Athletes who train their entire lives for a gold medal feel hollow the day after they win it. Founders who sell their companies can't get out of bed. The person gets what they wanted, the wanting disappears, and then they disappear with it. Over millions of years, evolution built us to struggle, to solve, to strive, to earn, and that wiring doesn't shut off just because our external problems go away.

When there's nothing left to fight for, the wiring turns inward and starts eating us alive. We feel it as emptiness, restlessness, depression - this sense that something critical is missing but we can't name what it is. AI is going to solve the external problems faster than anything in human history, and I believe that's actually the easy part. The hard part is what comes next - learning to generate purpose from the inside when the world stops demanding it from the outside.

When we want nothing, we get lost. When we have everything we wanted, we get lost just the same. We can only stay content in this relentless, almost ridiculous balance between the two - always chasing and never fully arriving. This need - this ache for something to do that plagues humanity is not something a book can fix. It has been written deep into the algorithms of our minds over millions of years of evolution. We cannot change it, so we must learn to thrive with it. Once we do, we will enter a true AI utopia. The good news is that you don't need to be a computer whiz or a philosopher or some kind of genius to figure this out. The answer isn't technical, for now, it's mostly philosophical. And it starts with understanding what purpose actually is, where it comes from, and how to build it yourself - even when the world is no longer forcing it on you. That's what Section 2 is about.

Section 2
The Science of Meaning

Chapter 5
The Ingredients of Purpose

To restate the thesis of the Purpose Problem: In order to have a meaningful life, you need purpose. Almost everyone gets purpose from one of two things: working toward something beyond themselves that they believe in, or working simply to survive. Both of these, as we know them, will be gone when ASI arrives. Without work, a mass loss of purpose seems like our destiny, yet no one seems to have a solution for how we can grapple with that as a society. That changes in this section.

The study of meaning is rooted in one central question, and that is: how do we fill the deep, dark abyss that lives within each of us? The more hollow we feel in a false AI utopia, the more desperately we need purpose and meaning to fill that void.

Many people think there is something uniquely special about "work" as we know it that makes it meaningful. And you can see why they'd think that - work gives you an identity, a routine, a community, a reason to get out of bed, and the feeling that you're contributing something to the world. But that is not work being special. That is work accidentally delivering the ingredients of purpose. According to Self-Determination Theory (SDT), developed by Richard Ryan and Edward Deci, a sense of motivation and purpose comes from three basic psychological needs: competence, which is mastery, achievement, and skill-building; autonomy, which is self-direction, choice, and authenticity; and relatedness, which is connection to others. A good job encompasses all of those.

SDT is a solid general framework for understanding purpose, but it is missing one key ingredient - the thing that makes purpose actually carry meaning. Researchers Frank Martela and Michael Steger studied what makes life feel meaningful. Martela, working directly with Ryan, one of SDT's founders, found that the three basic needs do generate meaningful experiences - experiences that feel comprehensible, directional, and significant - but that one crucial thing was still missing. Their research pointed to a fourth ingredient, which they call "beneficence," or doing good for others, as specifically necessary for creating a meaningful life. But beneficence doesn't fully capture it because I don't believe it's a fundamental requirement. I'd call it something more precise: transcendence. You can have competence, autonomy, and relatedness, but without being connected to and exerting your will on something beyond yourself, you will still feel lost.

The full recipe for purpose which includes all of these ingredients is what I call CART: competence - mastery, achievement, and skill-building, including the suffering, the wins, and the progress itself. Autonomy - self-direction, choice, and authenticity. Relatedness - connection to others. And transcendence - being connected to and exerting your will on something beyond yourself. You need all four.

In the chapters that follow, we will explore each ingredient of CART in depth. You will learn the science behind each one, what happens when it goes missing, what happens when you get too much of it, how to measure it, how you and society can actively cultivate it, and how it interacts with the other three. Understanding each of these ingredients intimately may very well be the difference between you feeling like a lost, hopeless, and empty soul and you thriving inside of an AI utopia. The following chapters will give you a roadmap for how to turn your false AI utopia into a true one. Let's get into it.

Chapter 6
Competence

When I refer to competence in this section, I am not simply referring to "being good at something," the way competence is typically defined. I am referring to something much greater - the art of mastery, suffering, achievement, and progress itself. The word "passion" originates from the Latin *passio*, meaning "to suffer" or "to endure." Your competence is both your passion and your *passio*. Building competence doesn't just mean you've unlocked a new skill. It means you have set a goal, gotten knocked on your face, picked yourself back up, planned a new strategy, executed it, and repeated that process until you achieved your goal. Then, having made that progress, you find a new goal - that is harder than the last - so you can repeat this process until you conquer that one too.

Anders Ericsson spent decades studying elite performers and proved that what separates experts from amateurs is truly not talent, it's "deliberate practice". This means setting goals just beyond your ability, failing, adjusting, and repeating until you succeed. Mihaly Csikszentmihalyi discovered that the "flow state" - a complete absorption in challenging work - only happens when difficulty matches skill level, right at the edge where you might fail but could win. Self-Determination Theory confirms that feeling competence through challenge and growth is a universal psychological need. Eliminate the competence cycle, and you lose the fundamental human drive to get better, which research proves is essential to purpose.

Howard Hughes set world aviation records, designed revolutionary aircraft, and built a billion-dollar empire. His life was defined by obsessive goal-setting - from breaking the transcontinental flight record, to building the largest airplane ever made, to turning TWA into a powerhouse. His competence cycle looked like this: he would set an impossible goal, fail repeatedly, figure it out, win, then find something even harder. After a brutal plane crash in 1946 and health issues, he stopped. Despite his riches, he withdrew from engineering and business entirely. By the 1960s, he was living on the top floor of a Vegas hotel, refusing to cut his hair or nails, unable to even set the goal of leaving his room. A man who once conquered the sky couldn't conquer a single day. With no aircraft to design and no records to break, he had nothing to obsess over, no failures to overcome, no wins to celebrate. He had conquered everything he set out to conquer. And once there was nothing left to conquer, he didn't know what to do with himself.

Ernest Hemingway is another example. He rewrote the ending of *A Farewell to Arms* thirty-nine times. He would sit down at a typewriter and bleed until he won. He would set a word count, fail to write anything good, tear it up, try again, finally nail a paragraph, then set the bar higher the next day. By the late 1950s, his body and mind had deteriorated. Even worse, he had lost his ability to write. He'd sit at the typewriter unable to produce anything that met his own standards. He had riches, fame, and a Nobel Prize, but without the daily struggle of failing at sentences and eventually getting them right, his life had lost its meaning. He couldn't finish his final book. He even stopped setting writing goals because he knew he couldn't complete them. Without that competence cycle, he fell apart. He had been defined by his competence as a writer, and so when he could no longer write, he could no longer be Hemingway. He told his friend A.E. Hotchner in their final conversation: "I can't finish the book. I can't. I've been at this damn table day after day after day after day." Three weeks later, Hemingway killed himself.

I should also mention that competence doesn't just build purpose on its own - it's also one of the fastest ways we build relatedness. In the 1950s, psychologist Muzafer Sherif and his wife Carolyn Sherif deliberately engineered real aggression between two groups of eleven- and twelve-year-old boys at a summer camp in Oklahoma. They split them into two groups - the Eagles and the Rattlers - and had them compete against each other until they genuinely hated each other. They were raiding cabins and burning flags to the point that the researchers had to physically separate them. Then the Sherifs introduced a challenge that neither group could solve alone. A truck carrying food for both groups had gotten stuck in the mud on the road back to camp. To eat, they had to work together to drag it out. They did, and almost immediately, the aggression began to dissolve. Shared challenges create very strong human bonds.

Naturally, most people derive their competence from work. They are given a task to do, which becomes their goal. They need to figure out how to achieve it. They will usually fail somewhere in the process, because their first approach isn't always right. When they do fail, they get back up and try again, and finally, if the job is the right level of challenge for them, they succeed. Without work, all of that disappears. There will be no revenue goals to chase, no team to lead and grow, and no professional identity to push forward. All gone.

When competence is missing, apathy loves to take its place. Your sense of forward progress stops, and so your life dulls. Everything you do is no longer about getting better, it's just about continuing to exist, which is an extraordinarily depressing way to live. You begin to forget why you even wake up in the morning if you're going to fall asleep the same person, not having improved one bit. Without challenge, there's nothing to obsess over, so you get stuck inside your own head, and without wins, life starts to feel like your enemy.

However, having too many competence cycles - too many challenges at once - pulls you in the opposite direction. If every day you're working to improve at twenty different things simultaneously, you'll feel overwhelmed and end up working with less focus because you won't know where to start. Like so many things in life, too little makes you depressed, too much makes you anxious. The goal is to maintain a healthy number of active competence cycles in order to keep yourself sharp and progressing, not so many that you're paralyzed.

So, how do you measure competence? The competence cycle looks different for every person, making it impossible to measure objectively at scale. But that doesn't mean you can't measure it for yourself. Every day, ask yourself these three questions and adjust your goals accordingly:

"Am I failing at something right now?"

YES → Good. You're challenging yourself.

NO → Warning. This is a depression zone.

"Do I believe I can eventually figure it out?"

YES → You're in the sweet spot.

NO → Warning. This is the anxiety zone.

"Am I obsessing over how to solve it?"

YES → Competence is active.

NO → Warning. It may be either too easy (boring) or too hard (you've already given up).

After working through these questions, you'll have enough clarity to assign yourself what I call a competence score. Rate yourself on a scale from 1 to 10: a 1 means you are never failing and everything feels easy; a 10 means you are always failing and everything feels impossible. If your answers were NO, YES, NO, you're in the 1-3 range. If your answers were YES, NO, NO, you're in the 8-10 range. If your answers were YES, YES, YES, you're in the 4–7 range. Record your competence score each day and make sure that over the course of each week, month, and year, your average stays within that 4–7 range.

On a societal level, measuring competence through business success rates won't work - there are too many edge cases and outliers, making it a poor predictor. And frankly, by the time this score becomes important, traditional companies will not even exist in the way we know them. The best approach will be to survey every individual in that society and ask for their yearly average competence score.

In a false AI utopia, if you personally fall below the optimal range - sitting in the 1–3 zone, which will be by far the more common symptom - you must take it upon yourself to add new challenges into your life. These challenges don't have to be transcendent. They don't need to involve relationships or increase your access to resources. They simply need to be hard enough to matter. Personally, if my score dropped too low, I'd start learning more advanced magic tricks, new moves on ice skates, or a harder chess opening. I'd try a new parkour course or try building a real friendship with someone I fundamentally disagreed with. Whatever. It doesn't matter if your challenges look nothing like mine - what matters is that they push you to the edge of failure without making you feel like progress is impossible.

Slowly fill your life with more of these challenges until you land inside the optimal range. Once you're there, you can direct your energy not only toward those new challenges, but toward the other ingredients of purpose as well. If you refuse to find new challenges and instead choose to spend your days consuming - believe me, you will become a very depressed, very unpleasant person.

Of course, knowing all of this and actually doing something about it are two very different things. Getting up and pursuing new challenges isn't easy, for two reasons. First, you will lie to yourself about where you actually are. When your only compass is how you feel in the moment, your emotions will rationalize your experience, convince you that you're fine, and keep you comfortable and sedated. Giving yourself a competence score forces your logical brain into the conversation, which makes it much harder to deceive yourself. Once you've done that, it becomes a very simple math equation: IF score < 4 THEN find challenge. Simple as that.

Second, challenges are often only enjoyable once you're in the middle of them or coming out the other side. The starting point doesn't always feel inspiring. That friction at the front end is one of the most common reasons people never begin. The competence score helps with the first problem. Nothing fully solves the second besides you. Inside a phase three society, there is no strategy I can hand you that will get you off your ass for you. That part is entirely on you.

Chapter 7
Autonomy

The word autonomy comes from two Greek words - autos, meaning self, and nomos, meaning law. You are your own law. Depending on how much of it you have, that's either the best thing in the world or the worst. A man with no choices is a prisoner. A man with infinite choices and nothing compelling him to make any of them is something worse. History knows the first man well, so let's start there.

Purpose, by its very nature, cannot be handed to you. The moment someone else selects it for you - even with the best intentions - it no longer holds any meaning. You might go through all the motions, challenges, wins, real community, and still feel an inexplicable emptiness. That emptiness occurs when the other three ingredients of purpose are present, but choice is absent. And even when you do fully choose your own goals and challenges, your own people to be around, and your own reason worth fighting for, it won't feel truly yours for long if you lack the freedom to change it, or the freedom to decide when and how to pursue it.

Autonomy is a unique ingredient because it is nearly useless on its own. Autonomy only gains real value when paired with the other ingredients of purpose - and the other ingredients only gain real value when paired with autonomy. So if you ever hope to find purpose, and meaning within it, autonomy is non-negotiable.

A 2024 meta-analysis of over one hundred studies by Vite, Patall, and Chen found that higher autonomy was linked to greater well-being, greater life satisfaction, and lower levels of depression and anxiety. Lower autonomy was also linked to more stress and burnout.

Consider the thousands of enslaved Black Americans who were master craftsmen - blacksmiths, carpenters, silversmiths, coopers, and weavers. Some were more capable than any free artisan within their region. However, even with extraordinary competence, they still reported high levels of meaninglessness and depression. This is because they had no autonomy - couldn't choose their projects, couldn't refuse work, couldn't quit, couldn't own their creations, couldn't benefit from their skill. A master blacksmith who forged perfect horseshoes and tools for forty years gained zero purpose from his competence because he never chose to do it. The competence cycle - set goal, fail, learn, master, harder goal - was replaced with: be ordered, comply, repeat. The only purpose they found through their work was in surviving.

For centuries, royal families and nobles across medieval Europe used forced monasticism to neutralize political rivals without killing them. They'd shave a rival's head, put him in a monastery, and force him to take vows. Some of these forced monks were highly educated and capable. The monastery gave them challenging work, a structured routine, and the chance to master theology, Latin, and manuscript illumination. But forced monks still became deeply depressed, tried repeatedly to escape, and some grew their hair back in an act of defiance. The reason for this was a lack of autonomy. They never chose a monastic life and so being compelled to pray, study, and serve - even when those same tasks gave voluntary monks profound purpose - felt like imprisonment. They had the same tasks, walls, and prayers, but one felt free while the other felt shackled.

The most fulfilling work in the world can feel like forced labor without autonomy. Strip it from any pursuit - no matter how meaningful it appears - and it becomes a cage. People who are never given the choice to do what they think is best, who are never asked for their input or allowed to act on their own judgment - they never get to feed what Nietzsche called the will to power. This breeds meaninglessness, and eventually, resentment.

For most of human history, the threat to autonomy ran in one direction - the danger was systems and people that crushed your freedom to choose. That threat will remain one for a long time, but inside a world transformed by AGI, a second and arguably more dangerous threat comes from the opposite direction. This is the problem the second man faces. The one that comes not from too many constraints, but from too much freedom.

Today, even the most autonomous person still has a floor. You still have to show up in some form, or the bills don't get paid. That constraint - as uncomfortable as it sometimes feels - is also a healthy guardrail. In a post-AGI world, that floor disappears. You no longer just have autonomy over what you do or how you do it - you have autonomy over whether you do anything at all. There is a point where freedom becomes so unconstrained that it produces the same emptiness as having none.

When a person can go an entire day, a week, even a month, without needing to produce anything, choose anything, or face anything - and faces no consequence for it - the ingredients of purpose begin to atrophy. This is what I call autonomy hyper-extension. And the solution is not to surrender your freedom. It's to wield it before it turns against you.

There is a concept in philosophy and law called the Ulysses contract. When Ulysses sailed past the Sirens - creatures whose song was so irresistible it drove sailors to leap from their ships - he didn't simply hope he'd be strong enough to resist. He used his present clarity and autonomy to bind his future self. He had his crew fill their ears with wax, ordered them to tie him to the mast, and made them swear to refuse to release him no matter how desperately he begged. His choice to be bound in the future was entirely autonomous. When the Sirens sang and Ulysses screamed to be freed, the ropes held and he survived.

This is exactly what you must do in a phase three world. Use your autonomy now, from a clear-headed place, to bind the version of you that will one day wake up in a world of infinite comfort and choose to do nothing. Preserve your freedom to choose what you do and how you do it, but remove your future freedom to opt out of doing something altogether. There is no dial to turn and no switch to flip that can do this. There is only your mind, and the decision you make right now about the kind of person you intend to be when the Sirens start singing.

Make a commitment to your future self that no matter how comfortable the world becomes, you will always search for something to become competent at, for people to become competent alongside, and for a reason worth being competent for. If it pushes you out of your comfort zone, if it feels completely against your instincts, if everything in you wants to sit still - search anyway. That commitment, made today from clarity, is not a limitation on your autonomy - it is the highest use of it. You are Ulysses. Tie yourself to the mast before you can hear the song.

At the same time - do not let anyone, including me, tell you what to find or where to look. Choose whatever challenge calls to you. Choose whoever you want beside you. Choose your own reason, and make sure it's actually yours - not something you inherited, accepted without questioning, or are still chasing because quitting would feel like losing.

This kind of self-imposed discipline - choosing to bind yourself to something in the future when nothing external forces you to - is one of the hardest things a human being can do alone. That is why the individual solution is insufficient on its own. For this to work at scale, society must evolve to meet people halfway. Ideally, a phase three society will culturally shame you for refusing to find or create a purpose.

Shame gets a bad reputation - and when it's pointed at the wrong things, it deserves it. But our brains have always used social approval and disapproval as a compass. Right now in the West, we shame poverty, unconventional paths, and failure. A well-built phase three society will point that same mechanism at purposelessness. And even after ASI, our mammalian brains that *despise* shame will still be there to influence our decisions, meaning we will effectively be nudged into choosing to do something even if we'd rather sit on the couch and consume.

An ideal phase three society would not push you toward any particular path - not medicine, not law, not anything specific at all. It would push you, relentlessly, toward following your own. In a culture where freedom and self-determination are celebrated, and running from your own will is seen as a failure of character, your autonomy balance will tend to correct itself naturally. When having a purpose becomes as socially expected as having a job, humanity takes one large step toward a true AI utopia.

If you see someone who has been handed too much autonomy and is drifting, do what you can to pull them back toward purpose. Nudge, challenge, shame if you must. Get them started on their quest. And if you see someone trapped beneath too little autonomy, suffocated by systems or people who won't let them choose, encourage them to break free and follow their own will. We are not meant to do this alone, and we are not meant to stand by while others get lost in either extreme of autonomy.

When the day comes that you no longer need to work and you find yourself debating whether finding a purpose is even worth the effort, know that you've been handed too much autonomy. Remember the commitment your past self made and allow your past self to strip away some of your present autonomy. And if you find yourself grinding toward something you no longer care about, alongside people who no longer excite you, for a reason you can't stand behind - know that you are not exercising the autonomy you should still have. You are in a prison. Break out, no matter the cost. Meaning tends to be waiting just past the bars.

Chapter 8
Relatedness

Once someone has chosen a competence that aligns with their will, they'll quickly feel full of purpose and energy. They'll wake up the next morning excited to start the day and begin their journey toward mastery. But that won't last long. Usually after only a couple of weeks - or less - they'll begin to feel empty inside again. At first, they won't understand why. They may think they've fallen out of love with their competence, but they'd be wrong. The real issue is that they're missing one of the most critical ingredients of purpose: relationships.

Relatedness, within the framework of purpose, is the who to the what and why you already have. It is the people you fail with, the people you grow alongside, the people you learn from, the people you teach, the people you show your achievements to, the people who acknowledge you throughout the process, and the people who care for you along the way.

For most of modern life, the majority of people's relationships have existed outside of work. Their spouse, their children, their childhood friends, their neighbors - all outside the workplace. And yet, we spend nearly a third of every day with our coworkers and supervisors - likely more than with friends and family members. We build more of our sense of connection and acknowledgment at work than almost anywhere else in our lives. When the job disappears, the paycheck can replaced - and it will be. The relatedness cannot.

Harvard's Study of Adult Development tracked hundreds of people over more than eight decades - measuring income, health, relationships, and just about everything else you could think of. The conclusion was that close relationships were the single strongest predictor of a long, happy life. It wasn't money, fame, or professional achievement. People who were most satisfied in their relationships at age fifty were the healthiest at age eighty. The study's director, psychiatrist Robert Waldinger, summarized it in one sentence: "Loneliness kills. It's as powerful as smoking or alcoholism."

Relatedness doesn't just make life feel better - it also keeps you from quitting when things get hard. A long study of nearly 1,700 young students in Norway found that loneliness was the single strongest predictor of students' intention to leave - stronger than grades, parental support, or teacher relationships. Just making one real friend in the first few weeks turned out to be one of the biggest factors for long-term persistence. Researchers at Yale found that people enjoyed chocolate more when someone else was eating it at the same time - even without talking. Bad chocolate tasted even worse when shared. The mere presence of another person amplifies the intensity of any experience, good or bad. Winning feels bigger when someone sees it and so does failing - but paradoxically, that shared sting after failure is also what makes people get back up and try again.

Marie and Pierre Curie discovered radium together. Marie could do science alone - she had done it for years before they met. But the partnership made breakthroughs possible that neither could have achieved solo, and more importantly, it made the work mean something more than it did when they were working alone. When they met in 1894, Marie was a brilliant but isolated researcher struggling for lab space and funding in Paris. Pierre was an accomplished physicist working on crystals. Together, they divided their labor in a way that multiplied both of their capacities. Marie performed the grueling chemical separations, processing eight tons of pitchblende to extract a single gram of radium, while Pierre conducted the precision measurements after each step. They worked side by side in a leaky shed for four years. Pierre had written to Marie before they married that he barely dared believe it possible - "to pass through life together hypnotized in our dreams: your dream for your country; our dream for humanity; our dream for science." In 1903, they shared the Nobel Prize in Physics.

Then, in 1906, Pierre was killed by a horse-drawn carriage. Marie was devastated. She wrote in her diary: "My Pierre, I think of you without end, my head is bursting with it and my reason is troubled... I no longer love the sun or the flowers. The sight of them makes me suffer." She continued their research alone. In 1911, she won a second Nobel Prize - this time in Chemistry for isolating pure radium. She was still achieving extraordinary things. Her competence had not diminished. Her autonomy had not been taken. Her transcendence - serving science and advancing human knowledge - was fully intact. But she described the work as hollow. Years later, she confessed: "I am working in the laboratory all day long. It is all I can do." She used work to distract herself from grief, and so her productivity continued, but the meaning she derived from it died with Pierre. Without her partner, her competence and autonomy became meaningless.

On May 29, 1953, Edmund Hillary and Tenzing Norgay became the first humans to reach the summit of Mount Everest, yet the achievement meant little to either of them without the other. Hillary was a tall, awkward New Zealand beekeeper. Norgay was a seasoned Sherpa who had attempted Everest six times before. During the 1953 British expedition, they were paired together and immediately clicked. The night before their final summit push, they barely slept in their tiny tent at 27,900 feet, sharing hot drinks and agreeing that if either turned back, both would. At 11:30 AM on that clear morning, they stood together at 29,035 feet.

Reporters pressed them afterward: "Who stepped on top first?" Hillary refused to answer for years. "We climbed as a team," was all he'd say. Norgay would later write: "We reached the summit almost together." The achievement wasn't merely standing on the world's highest point. It was standing there with someone who understood, with absolute intimacy, what it had taken to arrive. Hillary spent the rest of his life building schools and hospitals for Sherpa communities in Nepal through the Himalayan Trust. He once said, "I believe that of all the things I have done, there's no doubt in my mind that the most worthwhile have been the establishing of schools and hospitals." Standing on top of the world did give him fame, however standing there with Tenzing gave him meaning.

A job, whether you realize it or not, gives you all of this. You can ask your boss questions and receive guidance from someone further ahead than you.

You can teach and mentor the people you manage, which gives you the feeling of being needed. You can collaborate with peers on shared challenges, which creates belonging. You can fail or succeed as part of a team, which amplifies both the pain and the reward. And you can be acknowledged by the people around you, which tells your brain that you matter. People often don't quit because they're bad at their jobs - they quit because doing good work alone feels meaningless. Research has shown that workplace ostracism is actually more damaging than outright harassment. Kipling Williams, the psychologist who has studied ostracism for decades, explains why: it attacks everything at once - your sense of belonging, your self-esteem, your feeling of control, and your sense that you even matter. Being yelled at is terrible, but at least it acknowledges that you exist. Being ignored tells you that you don't matter at all.

In phase three - the abundant, post-AGI world I described in Chapter 2 where material needs are met but work as we know it has vanished - all of this will be gone. Your coworkers, bosses, and employees will vanish. The relatedness that millions of people never realized they were getting from their jobs will evaporate, and the emptiness that follows will be unlike anything most of them have felt before. It won't feel like a crisis. Instead it will feel like boredom, like restlessness, like something you can't quite name. But it will be the absence of relatedness, and if you don't recognize it for what it is, it will consume you.

In a world of advanced AI, you will be able to converse with artificial companions sophisticated enough to feel exactly like humans. They will remember your name, your history, your fears, and your humor. They will never judge you, never leave you, and never bore you. They will be available at 3 AM when no friend would answer the phone. They will say exactly what you need to hear, exactly when you need to hear it. And they will slowly, imperceptibly, replace the messy, inconvenient, sometimes painful reality of human relationships with something frictionless.

This is not my speculation. In 2025, researchers from MIT Media Lab and OpenAI studied nearly forty million ChatGPT interactions and ran a controlled trial tracking almost a thousand participants over four weeks. The patterns showed that the more people used the chatbot, the lonelier they got, the more they depended on it emotionally, and the less they talked to real people. The researchers couldn't prove the AI was causing the loneliness - it's possible that lonelier people were simply drawn to use it more. But regardless of which came first, the correlation was there. The more people leaned on the AI for connection, the less they reached for humans.

You feel lonely, so you talk to the AI. The AI is easier than a real person - no scheduling, no vulnerability, no risk of rejection. So you talk to it more. You socialize with real people less. You become lonelier. You talk to the AI even more. It's a downward spiral. I believe this is one of the greatest dangers of phase three - not because the technology doesn't do what we want, but because it actually does what we want too good. Real relationships are hard. They require patience, forgiveness, inconvenience, and the terrifying willingness to be known by someone who might not like what they see. AI companions will require none of that. And for millions of people, given the choice between something difficult and something easy, the easy option will win every time - even when the easy option is, in my opinion, terrible for human society. Society in phase three must build guardrails around how AI companionship is deployed - I don't believe this should be done by banning it, but by ensuring it supplements rather than replaces human connection. I'll address the specifics of how to do that when I discuss societal design in Section 3 - but the principle itself is non-negotiable. Any phase three technology that reduces real human connection, no matter how good it feels in the moment, is a threat to society - even if not to purpose - and must be treated as one.

To solve the relatedness problem in a phase three world, make sure for each pursuit you take on, you have someone you can look up to, someone who can look up to you, and peers who share your hunger for mastery and your appetite for competition. Some of your richest moments in an AI-abundant world will come from chasing hard things alongside people you genuinely care about - laughing, pushing each other, and competing all the way to the finish line. That kind of aliveness is available to you every single day in phase three if you build your relationships the right way.

But how do you know if you've built them the right way? In 1974, psychologist Robert Weiss proposed that all human relationships serve six distinct functions, which he called social provisions. Researchers turned these into a test that's been used and confirmed hundreds of times since.

The first is attachment. It's the feeling that someone truly knows you and actually gives a damn. Without it, you can be in a room full of people and still feel completely alone.

The second is social integration - belonging to a group that shares your interests, values, and way of seeing the world. This is not being accepted, but being truly and deeply understood.

The third is reassurance of worth - people whose opinion you respect telling you that you're good at what you do and that it matters.

The fourth is a reliable alliance - knowing that if everything fell apart tomorrow, people would actually show up for you.

The fifth is guidance - access to people wiser or more experienced than you who can help you make hard decisions without pushing their own agenda.

The sixth provision is the one I think is most often overlooked, it's the opportunity for nurturance. This is the feeling that others need you. That someone else's growth or safety or success depends on you being there. This is what makes mentorship meaningful, what makes parenting one of the most purpose-dense experiences a person can have, and what makes leaders describe their teams as the best part of their work. It is also a form of transcendence, although not the only form as we will discuss in the next chapter.

If you're missing even one of these six, you have a gap in your relatedness that no amount of competence, autonomy, or transcendence can fill. Your job is to find someone who can offer you that provision and offer it back. Relationships are not one-directional. They cannot be. The moment you stop giving, you stop receiving - and the provision collapses. The mentor-peer-mentee structure I described earlier will cover all of these - guidance from those ahead, nurturance from those behind, attachment, reliable alliance, social integration, and reassurance of worth from those beside you.

With all this said, I should also warn you that it is entirely possible to have too much relatedness, and the consequences are just as destructive as having too little.

In April 1961, President John F. Kennedy authorized the Bay of Pigs invasion - a plan to overthrow Fidel Castro by landing 1,400 CIA-trained Cuban exiles on a beach in Cuba. The plan was, by almost every objective measure, terrible. The landing site was surrounded by swamps. The exiles were outnumbered. The assumption that the Cuban people would rise up and join the invasion had no intelligence to support it. And yet, Kennedy's inner circle - some of the most intelligent and experienced advisors in American government - unanimously approved it. Not one person in the room raised a serious objection. Afterward, Kennedy was stunned. "How could I have been so stupid?" he asked. The answer, as psychologist Irving Janis later demonstrated, was groupthink. The advisors were so tightly bonded, so invested in the group, and so afraid of being the one to break consensus that they silenced their own doubts. They had too much relatedness and not enough autonomy. The social pressure to conform was so strong that opposing opinions were never voiced, and the result was one of the worst foreign policy blunders in American history.

Over-relatedness also breeds a fixation on what everyone around you is achieving. When you constantly measure yourself against others, perfectionism and anxiety follow. The more people you're surrounded by, the more targets for comparison exist, and the harder it becomes to stay focused on your own competence cycle. And then there's codependency - the sneakiest and most dangerous form of too much relatedness. Codependency is what happens when you form such a bond that you lose the ability to function without the other person. This is the couple who can't spend a single evening apart. The friend who won't start anything unless the group approves first. The parent who poured every ounce of identity into their children and then collapsed into depression the day those children left for college. Codependency is the slow death of autonomy. You stop making your own decisions, you stop trusting your own judgment, and you lose who you are when the other person isn't in the room.

Too little relatedness makes life feel hollow. Too much makes you lose yourself. The goal is the same as it is with every ingredient of purpose - find the balance, protect it fiercely, and adjust when you notice yourself drifting toward either extreme.

Before I move onto the next chapter, I must leave you with the brutal truth of relatedness. Most people will not do any of this. Most people will take the path of least resistance, which in phase three means retreating into AI companions, virtual reality, and the numbness of simulated connection. Real relationships are hard and simulated ones are easy. The gap between those two options will be wider in phase three than it has ever been in human history.

Start now. Before phase three arrives, before the Sirens of simulated companionship begin to sing, build real relationships that will anchor you when everything else changes. Make sure the people around you are real, that they challenge you, that they need you, and that they'll still be there when the frictionless alternative shows up at their door. Relatedness is the most portable ingredient of purpose - there are good people everywhere, in every city, in every community, in every pursuit worth having. But it is also the most fragile. The Curies had it and lost it to tragedy. Millions of workers have it right now and will lose it to automation. You have it in some form, and you will lose it too if you don't protect it with the same urgency you protect your health, your freedom, and your will. Do not let the most human part of purpose be the first thing you surrender to a machine.

Chapter 9
Transcendence

I f you choose a goal to strive towards, follow your will on your quest to achieve it, and you do so with people that you love - you will still feel an emptiness in your heart that you will not know how to fill. To close this final gap, you need to transcend. I don't necessarily mean that in a spiritual way, although I will come back to that. Instead, by "transcendence," I simply mean: for something beyond oneself. Even if you have found your purpose through choosing a goal, working towards it, and surrounding yourself with the right people - it will still remain a hollow and useless purpose if it isn't rooted in a reason beyond yourself. That transcendent reason can be for other people, for animals, for the earth itself, for God - hell, even for aliens. As long as it is for something beyond you.

Nietzsche once wrote: "If we have our own why of life, we shall get along with almost any how." Viktor Frankl, drawing from his experiences in Nazi concentration camps and his decades of clinical practice, proved him right. Frankl noticed what he called the "existential vacuum" - a crisis of meaninglessness that shows itself as boredom, apathy, depression, aggression, and addiction. When Frankl saw that unemployed young people were showing signs of depression and meaninglessness, he told them to volunteer at a library or youth group. Through these activities, the young people quickly found meaning and no longer showed signs of depression even though they were still broke. This is because they began working for something beyond themselves.

Frankl's own survival in Auschwitz is one of the most powerful examples of this. In the concentration camps, he observed that the prisoners who survived longest were not always the youngest, strongest, or healthiest - they were those

who maintained a reason beyond themselves. Frankl himself survived by mentally "giving lectures" to imaginary audiences about the psychology of camp life,
turning what he learned from his suffering into a gift for future generations.
He witnessed other prisoners cling to life in order to reunite with loved ones,
to witness the atrocities so the world would know, or to complete unfinished
work. One man held on because he was writing a scientific manuscript that
only he could complete. Another survived to see his daughter again. A prisoner
in Auschwitz who had a goal, who wanted to survive, and who had comrades
around him - that man would still probably not survive without transcendence.
Those who lost their transcendent "why" - those whose families had passed
away, whose life's work had been destroyed, and who saw no purpose beyond
survival itself - often died within days of giving up. These men perished not from
starvation, but from meaninglessness.

Erik Erikson shows the same with his concept of generativity - the drive
to contribute to and nurture the next generation. Erikson found that people
who don't develop generativity by midlife tend toward self-centeredness, lack
of productivity, and deep feelings of disconnection. Harvard also found that
men who achieved higher levels of generativity at midlife showed much stronger
cognitive functioning and lower rates of depression three to four decades later,
even after controlling for intelligence and education. Generativity is only one
form of transcendence, but the pattern it helps to reveal applies to all of them.
When you stop living for something beyond yourself, some part of you begins
to die.

At fifty, Leo Tolstoy - already the praised author of War and Peace and Anna
Karenina - fell into suicidal depression even though he still had the other three
ingredients of purpose. He was extraordinarily competent and was at the height
of his literary powers. He had complete autonomy since he owned a vast estate,
controlled his time, and answered to no one. And he had deep relatedness, being
married to Sophia with thirteen children, surrounded by family and admirers.
Yet Tolstoy wrote in his Confession: "I could give no reasonable meaning to
any single action or to my whole life." He carried a cord out of the room
where he undressed alone every evening so he would not hang himself from

the beam, and he stopped going out to hunt with a gun so he would not be tempted to end his life. As he wrote: "I could not patiently await the end. The horror of the darkness was too great." His great novels had been written to help humanity, but he had reached a point where even that no longer felt like it was enough because he knew his writing would eventually be forgotten. He described himself as a man lost in a forest who was horrified after having lost his way, rushing to try to find the road, knowing that every step was leading him deeper into confusion - but unable to stop rushing. Tolstoy's suicidal thoughts went away once he rediscovered transcendence in what he believed to be an eternal form. After committing to Christian anarchism, renouncing his copyrights, working alongside peasants, and devoting himself to writing moral philosophy for humanity's spiritual development - he finally found meaning in his purpose through transcendence.

When you lose your transcendence, you are at the beginning of the end, and we see that over and over again. Now, it's going to get worse. If you ask almost anyone what their dream job is, I'd wager they say something along the lines of: "Oh, I'm different from most people because I really just want to help others with..." and then they name something beyond themselves. In a pre-AI utopia, transcendence is relatively easy to cultivate. All you have to do is choose a target - children, veterans, the homeless, the impoverished, rainforests, endangered animals - and find a way to structure your competence, autonomy, and relatedness around serving them. In an AI utopian world, we will no longer need to worry about that in the same way. We won't get to derive meaning from keeping people alive or providing a great life for them. There will no longer be urgent problems we need to solve, but our desire to solve them will not go away. The world will not need our help to save itself, but we will still want to help it.

The lack of transcendence that this will create is going to breed a deep sense of meaninglessness. A feeling - felt by everyone who used to work for something beyond themselves - that they are no longer making an impact on anything outside themselves. Angst will begin eating transcendence-less people alive from the inside, forcing them to keep asking: Why does my life even matter? People who used to bury this question under the stack of papers on their desk - hoping to never confront it - will no longer be able to hide. They will have to face it, over and over again.

So how do you know how much transcendence you even have? Most people assume they know, but if they aren't honestly interrogating their own lives, they don't. Start by asking yourself whether you genuinely care for something beyond yourself - not in the way that everyone "cares about the world," but truly - in a way that actually changes how you spend your time. Then ask whether you are actively working toward something beyond yourself, or merely thinking about it. And finally, honestly look at how much of what you do on a daily basis is for you, and how much is for something or someone else. If you answer those questions truthfully, you'll know exactly where you stand. The more you do for something beyond yourself, the more you are truly transcending.

If after asking yourself these questions you discover you're not living a large part of your life for something beyond yourself, you need to find something bigger than you worth living for. It could be spirituality, your children, a creative art form, or something as simple as a friend. It doesn't matter what it is as long as you care about it, are willing to work for it, and want to show up for it daily.

I predict that one of the biggest sources of this transcendence in an AI utopia will be helping others find their own purpose - including transcendence itself. We are already seeing this happen today. Even in 2025, one of the most common ways people derive transcendence is by helping other people find meaning. Motivational speakers, authors (yes, guilty as charged), podcasters, psychologists - they all do this. In this way, the world has become one giant meaning pyramid scheme. Older people derive meaning from helping younger people find meaning, which those younger people eventually use to get to a place in life where they can derive meaning from helping even younger people find meaning.

This pyramid - as much as it makes us look like ants - is so deeply ingrained in us because it has driven us to invest in the next generation across tens of thousands of years. It is, honestly, one of the only reasons we are still here. Think about it. The thing that kept our species alive was not raw strength or speed - we are weak and slow compared to most predators. What kept us alive was that generation after generation, people found enough meaning in helping the next generation survive that they kept going. They kept teaching, kept protecting, and kept sacrificing. The meaning pyramid scheme of human nature is, in many ways, the engine of human survival. And in an AI utopia, when the material reasons to invest in the next generation are handled by machines, the meaning-based reason to invest in them will be more important than it has ever been.

If there is a creator, it's a fascinating design choice to make a human incapable of achieving his own meaningful purpose without first either finding his creator or helping a fellow human on their path to purpose. It's a beautiful cycle, and one that will only grow more vital as we enter an AI utopia. Many people will have competence, autonomy, and relatedness - but they will fill the last and deepest hole in their heart simply by helping others achieve those same things. And so the cycle continues.

Now, it might seem like too much transcendence is impossible. How could anyone care too much about something beyond themselves? But they can, and the consequences of that extreme are also devastating.

In 1909, Simone Weil who was a French philosopher, mystic, and political activist scored first in the entrance exam at the Ecole Normale Superieure, beating out Simone de Beauvoir for the top spot. She had autonomy, competence, and connections, but from childhood, Weil's transcendence consumed everything else. At five years old, she refused sugar because French soldiers at the front during World War I had none. As an adult, she left her teaching career to work anonymously in a Renault factory so she could experience the suffering of the working class firsthand. She joined the front lines in the Spanish Civil War despite having no military training. And then, in 1943, while working for the French Resistance in London, she was diagnosed with tuberculosis and told to rest and eat. She refused. She limited her food intake to what she believed the residents of occupied France were eating - and likely ate even less than that because she could not bring herself to nourish her own body while others starved. On August 24, 1943, Simone Weil died of cardiac failure at the age of thirty-four. Her biographer Richard Rees wrote that whatever explanation one gives for her death, it amounts to saying that she died of love.

That is too much transcendence. Weil cared so deeply about something beyond herself that she destroyed herself in the process. Her competence atrophied because she could not work. Her autonomy was lost in some ways because she could no longer make rational choices about her own survival. Her relatedness also vanished because she isolated herself in a hospital bed. Transcendence, when it becomes all-consuming, eats the other three ingredients alive. And you don't have to starve yourself to death for this to happen. You see it in the activist who burns out completely because they can't stop fighting even when their body is breaking and you see it in the religious devotee who gives away everything - money, time, relationships - in service of a cause, and wakes up one day with nothing left. Transcendence without the other ingredients is another form of suffering.

In an AI utopia, the risk of Weil-level transcendence will be lower than it is today, simply because our basic needs will be met regardless. You won't starve yourself to death in solidarity with others because no one will be starving. But the subtler forms - losing yourself entirely in a cause, neglecting your own growth and autonomy in service of something bigger - those will persist. And they will remain dangerous. However, they will continue to be much less of a concern than too little transcendence.

To be clear, transcendence is not to be confused with relatedness. Relatedness is who you do things with. Transcendence is what or who you do them for. Finding transcendence ultimately comes down to doing what you do for God, for your fellow humans, or for the other living creatures of this world. It's exerting your will on something that is larger than just you. That's it. My recommendation is to find something you can do for your fellow humans that no AI can replicate. Then do that thing as much as you possibly can - because that is an irreplaceable transcendence. And it could be as simple as being the one who makes your friends laugh. Or it could be as vast as helping the next generation find their own reason to keep going, so they can one day do the same for the generation after them - the pyramid doesn't need to stop. Remember, in an AI utopia, the whys aren't gone, they're just waiting to be found.

Chapter 10
The CART Recipe

If you've made it this far, you know CART: competence, autonomy, relatedness, and transcendence. You know what each one means, you know why each one matters. But knowing the ingredients of a recipe and knowing how to cook with them are completely different things. You can stare at flour, eggs, butter, and sugar on a counter and understand, intellectually, that they can become a cake. But if you dump them all in a bowl without understanding their relationships, you end up with an inedible mess.

Purpose works the same way. You need all four ingredients because without one, nothing works. And you need to understand how they work together, how they clash, how they reinforce one another, and most importantly, how to combine them in a way that makes you feel truly purposeful.

Take a moment to think about making your friend laugh over dinner. It's a small moment - it's not going to make the history books. Just you, across a table, telling a joke. You see an opening when your friend tells a story about their boss, and you realize there's a perfect callback to something that happened last week. You choose to make the joke right then and there, and you don't do it because you're performing for an audience or because anyone is forcing you, but because you want to. That's autonomy. You time it just right. You pause before the punchline and you deliver it with the exact tone and inflection that makes your friend burst into laughter. You've done this before, and you keep getting better at it. That's competence. Your friend is with you in this moment and so the joke means more to them because it comes from you. When they laugh, you laugh. When their eyes light up, yours do too. That's relatedness. And you're not telling this joke to prove how clever you are. You're telling it because you want to bring a little joy into your friend's day - you're doing it for them. That's transcendence.

In that moment, you don't just feel happy. You have all four ingredients and so you feel purposeful. Like you're doing exactly what you're supposed to be doing, exactly where you're supposed to be. I think most people have felt this at some point in their lives - maybe not at a dinner table, maybe somewhere else entirely - but they've felt it. The problem is that most people feel it randomly, by accident, and then have no idea how to get it back. That's what this chapter is for. By the end of it, you'll know exactly how to create that feeling on purpose, and more importantly, how to keep it from falling apart.

Imagine if you could live your entire life structured this way. If from morning until night you could work on something you're genuinely good at, that you chose because it matters to you, alongside people who see you and support you, in service of something greater than yourself. Most people don't have all four and when you're missing even a single ingredient, the entire CART recipe falls apart. Here's what happens when you have three out of four:

John Kennedy Toole had everything except proof that he was good enough. He had autonomy - he wrote what he wanted to write, a comedic novel called A Confederacy of Dunces that captured the absurdity of modern life and the craziness of New Orleans. He had relatedness - he was close with his mother, had friendships, and felt deeply connected to the culture he was writing about. He had transcendence because he wasn't writing for fame or money, but to preserve a disappearing world and to create something that would outlast him. But he couldn't get published. Every editor rejected him and every single door he tried to open slammed shut. Now, technically, Toole was competent - the Pulitzer Prize his book later won proves that. But competence as an ingredient of purpose isn't just about getting and being good at something, it's about knowing you're getting better and already good at something. It's the full cycle I talked about in the competence chapter· struggle, fail, learn, win, then set a harder goal. Toole never got the win. He had the skill but the cycle was broken, and without the world telling him "yes, this was worth it," competence couldn't do its job as an ingredient. It just sat inside him, completely useless. And without that feeling of competence, the other three ingredients began to crumble. His autonomy felt pointless, his relationships felt hollow, and his transcendent purpose felt delusional - because you can't preserve a world through a manuscript that sits in a drawer. At thirty-one years old, he drove to a remote area outside New Orleans, ran a hose from his car's exhaust pipe into the cabin, and ended his life.

Sylvia Plath had the proof of competence that Toole died without. She was a brilliant poet who was published, awarded, and recognized by the literary world while she was still alive. She had relatedness, being she was married to Ted Hughes, had two children, and was deeply embedded in literary circles. She also had transcendence - she was writing confessional poetry to give voice to women's inner lives, to articulate experiences that an entire generation of women had been forced to keep silent about. But she didn't have autonomy. It was the early 1960s and the world expected her to be a perfect wife, a perfect mother, and a perfect poet all at once, all without complaint. Her husband's infidelity ruined the domestic life she was told by everyone she had to maintain, and she couldn't leave without being blamed for destroying her own family. However, she also couldn't stay without dying inside. Her mornings were spent making meals and changing diapers. Her poetry - the thing that was keeping her alive - had to be written between four and eight a.m., before the children woke, in a freezing London flat during one of the coldest winters England had ever seen. Every hour of her day felt like it belonged to someone else. She had skill, she had connection, and she had purpose - but every one of those ingredients was trapped behind a wall of obligations she never chose and couldn't escape. At thirty, she sealed the kitchen, turned on the gas, and put her head in the oven.

Three out of four ingredients is not enough. It will NEVER be enough. You can have an extraordinary amount of three ingredients and still feel like your life is falling apart, because the missing one creates a hole that the other three cannot fill no matter how strong they are. You cannot substitute one ingredient for another. Extra relatedness doesn't make up for missing competence. More transcendence doesn't replace autonomy. They multiply each other when present, and they pretty much all collapse on each other when one goes missing.

Before moving on, I want to be very clear about something. You do need all four in your life, but not in every single moment. You can think of these four ingredients just like you think of nutrition - you don't need protein, carbs, fats, and vitamins in every bite, but over the course of a day you need all of them. If you spend a morning alone in deep focus, you're exercising competence and autonomy. Relatedness and transcendence aren't present and that's fine. After an evening with friends and a call from someone who looks up to you for advice, you have exercised them too. Similarly in nutrition, a meal of only carbs may not make you feel your best, but it's okay. However, go a week eating only carbs and you'll really feel it. Go too long without one of the ingredients of purpose, and believe me, you will feel the emptiness creep in even if you can't name what it is.

Now, it's important for you to know that there is a problem that may arise as you begin to build your life around the four ingredients. Sometimes the ingredients don't complement each other - they actually fight each other. You're great at law, but you want to be a musician. Your friends want you to stay in town, but you need to move across the country. You could make millions in finance, but teaching serves humanity better. If you don't have a way to resolve these issues, you will be stuck forever - you'll feel like every direction is a betrayal of something important. When your ingredients are pulling you in two directions at once, there's a hierarchy you can follow. This hierarchy exists because not all four ingredients of CART are equally foundational, and when they fight each other, one has to win.

Transcendence is the why. Without it, everything else is just busy work to distract yourself from the void. This doesn't mean you should start your search for purpose here - you shouldn't spend your days serving God or humanity doing something you suck at or hate. However, it's better to live your days doing that than live your days serving nothing at all. Transcendence is the ingredient that can turn any suffering into purpose if need be. When it conflicts with the others, you must always ensure it wins.

You could make millions in finance, on your own terms, with a great team, but if the work serves nothing beyond your own bank account, you'll feel empty. Better to take a pay cut, teach economics to underserved students, and feel like your life matters. Or even better, dissolve the conflict by keeping your finance job but using the money to help others after work instead of just partying with your friends.

Competence is truly your foundation. Without it, autonomy is just flailing around with no clear path, relatedness is just hanging out, and transcendence is valuable but it can't be acted upon in the same way. You can't serve humanity nearly as well as you should be able to if you're bad at everything and refuse to work towards a goal. You may still be able to serve God, but I'd argue not to the same extent. When competence conflicts with autonomy or relatedness, competence must win.

Autonomy is your will. Without it, competence is slavery. You may be good at something and working hard to get better, but it's draining you because it's not yours. When it conflicts with relatedness, autonomy must win. "I want to move to Bali to build my startup, but my family wants me to stay home" - if you sacrifice your autonomy for relatedness, staying somewhere you don't want to be, doing work you don't believe in just to keep other people comfortable, you'll resent them and resent yourself. Eventually the relatedness that you sacrificed everything for will rot from the inside out, poisoned by deeply rooted resentment. Besides, the best and deepest relationships survive autonomy. The people who actually love you will still be there when you follow your own path - and if they won't, that tells you something important about the relationship.

Relatedness is the most negotiable ingredient because it's the most portable. There are good people everywhere. You'll find your people in San Francisco, New York, Berlin, Bali, or anywhere else. There's communities online, at a coffee shop, at a writing retreat, everywhere. The other three ingredients cannot be found elsewhere because there is no elsewhere - they are within you. Transcendence, competence, autonomy, and then relatedness - that is the hierarchy whenever you find that ingredients are fighting each other.

Now, once you are in a place where you have all of the ingredients, you'll notice that they begin to feed and play off of each other. Getting better at something gives you confidence, and confidence gives you the courage to make your own choices. Having the freedom to choose what you work on means you choose things you care about, and when you care, you practice harder and improve faster. Being around people who believe in something bigger than themselves shows you purposes you'd never discover alone - maybe your friend mentions a cause, and suddenly you see how your skills could serve it. And working toward something meaningful attracts people who care about the same things. Achieving any single ingredient creates a positive feedback loop that helps you get the others.

Julia Child is a great example of this. In 1948, she was a thirty-six year old diplomatic wife who had never really cooked, following her husband Paul to a State Department posting in Paris. She had no career, no direction, no particular skill worth mentioning, he didn't know what a shallot was - she was just a tall, awkward American woman with no idea what to do with her days.

Then she ate sole meunière at a restaurant in Rouen on the drive into Paris and for some reason, something clicked. She enrolled at Le Cordon Bleu which was not in the beginner housewife course, but in the rigorous program for professional chefs. That was her competence clicking into place. She wasn't good yet, but she was working at getting better, and the work itself lit her up. And because she chose it freely - nobody told her to become a chef, her husband just encouraged her to follow whatever she wanted - the competence was paired with autonomy from day one. The competence then led her to people. She joined a cooking club, met Simone Beck and Louisette Bertholle, and the three of them started teaching cooking to American women out of Child's Paris kitchen. For the next decade, as she and Paul moved from Paris to Marseilles to Bonn to Oslo to Cambridge, she and her collaborators tested and retested recipes, writing hundreds of letters back and forth. She told her sister-in-law: "Really, the more I cook the more I like to cook. To think it has taken me 40 yrs. to find my true passion."

And the transcendence showed up last. What started as a personal fascination turned into a mission to make French cooking accessible to every American household. The book they spent a decade writing - Mastering the Art of French Cooking - changed how the entire country ate. The TV show that followed changed how the entire country thought about food. She went from being a bored diplomatic wife to serving millions of people who had never been taught that cooking could be joyful, that food could be fun, and that a home kitchen could be a source of real pride. The "why" showed up after everything else was already in motion, yet it was still the most important part of it all.

That's the upward spiral. One ingredient pulls in the next, each one makes the others stronger, and the whole thing accelerates. I think that's actually one of the most beautiful things about purpose - once you get it going, it kind of takes care of itself.

However, you must know that the spiral runs in reverse too, and actually even faster. I already talked about Howard Hughes in the transcendence chapter, but he's worth revisiting here because he's the clearest example of what happens when purpose unravels itself. In his twenties and thirties, Hughes had CART. He had competence (he was a groundbreaking film director, a record-setting aviator, and a brilliant engineer who designed aircraft that pushed the limits of what was physically possible). He had autonomy (he inherited a fortune that gave him total freedom to pursue whatever fascinated him). He had relatedness (he was surrounded by col-laborators, friends, and some of the most famous women in Hollywood). And he had transcendence (he genuinely believed he was advancing avia-tion for humanity, pushing the frontier of what humans could do).

Then, slowly, he started losing one ingredient. His OCD - which had been there since childhood - tightened its grip on his mind. As it worsened and he wasted more hours on rituals, he started pulling away from people slowly - fewer meetings, fewer dinners, more hours alone in darkened rooms. Relatedness was the first ingredient to go. And without people around him to challenge him, ground him, or hold him accountable, his autonomy became warped and confusing. Freedom without connection became isolation which became para-noia. He stopped trusting his employees; really, he stopped trusting anyone. His competence - that brilliant, restless mind that had designed aircraft and directed films - had nothing left to work on because he couldn't engage with the world long enough to build anything. And without competence, relatedness, or real autonomy, his transcendence collapsed. There was no larger purpose anymore, there was only fear.

By his final years, Howard Hughes was the richest man in America and arguably the most miserable person in it. He lived in blacked-out hotel rooms. He stopped cutting his hair, his nails. He stopped bathing. He was addicted to codeine, barely coherent, and completely alone. When he died in 1976 at seventy years old, his body was so emaciated that observers compared him to a prisoner of war. The man who once had every single ingredient had lost them one by one, each loss accelerating the next, until there was nothing left. That should terrify you. Because the difference between Hughes in his thirties and Hughes in his sixties wasn't just one single catastrophic event - it was really a thousand small surrenders that compounded until they became irreversible. If you keep failing at something without wins and you stop trusting yourself, if you defer to others instead of deciding for yourself, if you force yourself to work on things you don't care about and you stop improving, if you isolate yourself and let your "why" warp inward until it bears no resemblance to reality - once you're in that spiral, it's incredibly hard to escape. Make sure you start in an upwards one. And if you find yourself in a downward one, catch it early, name it, and fix whichever ingredient broke first before it takes the rest down with it.

So, where do you start? Oddly enough, the best place to start is not at transcendence even though it is the most important ingredient. This is because, as powerful as transcendence is by itself, when it's paired with a real goal from the ingredient of competence, it becomes that much more powerful. And to find the right relationships, you need to have a common competence - that's another reason why competence is the first thing to think about on your quest for purpose. Julia Child didn't set out to change how America eats. She set out to learn how to cook. The mission came later, after the skill and the people were already in place.

So, let me ask you: What do you enjoy getting better at? What problems keep pulling you in? What makes you lose track of time? Don't ask what's meaningful yet - just ask: what am I good at, or what do I love that I could become good at if

I practiced? These questions help you find the thing that will provide you with competence.

Once you've identified it, ask whether it's actually yours. Are you pursuing law because you love it, or because your parents expect it? Are you building this startup because you believe in the idea and want to see it through, or because you think it'll impress the people who are trying to control your life for you? If the answer is the latter, stop. Every year you spend building someone else's version of your life is a year you don't get back, and the further you go down that road, the harder it becomes to turn around.

Then find your people. Who's already doing what you want to do? Who's a few steps ahead? Who is a few steps behind? Who can challenge you, work alongside you, and be honest with you when you're wrong?

Last, ask why it matters. Who does this serve? What does it protect? What does it leave behind? Maybe it's for your family, a community, or for a specific cause. Maybe it's just to make your friends laugh. Doesn't matter what it is as long as it is outside yourself and makes life feel like part of a story bigger than you can comprehend.

This brings us back to the start of this book. In a world where AI solves all material problems - where you don't need competence to survive, where every need is met, where social connection can be automated and suffering is largely eliminated - these four ingredients don't become less important. They actually become the only things that matter in society. There won't be jobs to hand you a default competence. There won't be offices to hand you a default community. There won't be bills forcing you to do something, anything, with your day. You'll have to build all of it from scratch. Julia Child found all four because she stumbled into competence first and let the rest follow. Howard Hughes lost all four because he let relatedness slip and the rest collapsed.

You won't need competence to eat, but you will need it to feel alive. You won't need autonomy to function, but you will need it to feel free. You won't need relatedness to survive, but you will need it to feel human. And you won't need transcendence to exist - but without it, you'll spend every day wondering why you bother existing at all.

Section 3

The True AI Utopia

Chapter 11

Building a World Worth Waking Up For

Once society no longer needs to survive or work, the principles of our current society will no longer suffice. In fact, they will do quite the opposite. Right now, society is built to reward you for following the rules, doing your job, and buying things. That entire system assumes you need to work and need to spend. In a true AI utopia, both of those assumptions are gone. People will not need to work and they will be able to buy whatever they want in exchange for nothing.

When that happens, the only thing left with any real value is meaning. Purpose will be the only thing people need to optimize for and so our society needs to be built in a way that supports that. As we saw in section two, in order to create a society that supports purpose, we need to create a society that supports CART.

The problem is that humans are really bad at optimizing for purpose when survival is already taken care of. Look at lottery winners, trust fund kids, retirees - if you give someone total security, you will often end up watching them fall apart. This is not because they, as individuals, are weak. It is because we're wired to need struggle. We're wired to need autonomy. We're wired to need connection. We're wired to need transcendence. If the necessity that forces us to pursue these things is taken away, most of us naturally just... freeze.

So we need to redesign society to create the right kind of pressure - this time not survival pressure, but purpose pressure. We need to make it so that living a meaningful life is the path of least resistance, and living an empty life requires actively fighting against the grain of society. Here's how we might do that for each ingredient.

I want to be upfront - the ideas I mention later in this chapter are going to make some people angry. I'm not claiming to have the one and only blueprint for a perfect society. But I'd rather put specific ideas on the table and be off with a few of them than do what most people do when they talk about the future, which is say "someone should figure this out" and leave it at that.

Transcendence:

For transcendence, our society needs to be set up in a way where people are living a large part of their everyday lives for something beyond themselves. Right now, in the current world, most people get their transcendence accidentally. They go to work, and their work happens to serve other people. They raise kids, and their kids happen to give them a reason to keep going. They don't sit down and think "I need transcendence today." It just shows up because the structure of life delivers it to them without them asking for it.

In an AI utopia, that structure is gone. Nobody has to work, nobody has to provide for anyone, and the things that used to accidentally connect you to something larger than yourself have all been taken care of by Divus. So transcendence has to become deliberate. People need to actively choose to live for something beyond themselves, and society needs to make that choice as easy and natural as possible.

The first most obvious path to transcendence is God. For people of faith, an AI utopia could actually deepen their relationship with God because for the first time in human history, they'll have the time and freedom to really pursue it. They will no longer be squeezing prayer in between shifts or choosing between church and overtime. I think religion will remain one of the most powerful sources of transcendence in phase three, but I also don't think there is much more religious people can do to push society toward it. People will either believe or they won't, and I think ASI will divide people even further - with one side thinking it proves the absence of God and the other saying it proves the truth.

But transcendence doesn't have to be religious. It can be contributing to your community, creating something beautiful for others, raising children, inventing a new game that thousands of people will love playing, or even just being the person who makes your friends laugh. The key is that it has to be for something beyond you, not just for you.

In society, talking about transcendence in a truthful way is sort of frowned upon. You are supposed to say you just want to help people, but you aren't supposed to say that you will feel empty and meaningless if you don't work for something outside yourself which happens to be people right now. I think it's time that changed. I think it's time society becomes more open about their true deep needs and is clear and open about them. We need to admit that altruism isn't purely altruistic - it's also selfish, because it fills a hole inside us that nothing else can fill. And that's okay. It's actually beautiful that we still help each other even though we are no angels. It means we're wired to need each other and wired to contribute to one another.

So what does society actually need to do? First, it needs to create visible, accessible ways for people to contribute. This will not be through traditional work, but through creation, community building, mentorship, and helping others in small, all too human ways. I believe every neighborhood should have what I'll call contribution boards - physical and digital spaces where people can post what they need help with and what they're willing to help with. Maybe someone wants a painting partner who will push them to get better. Maybe an older person wants to teach someone their native language. Maybe a group of kids needs an adult to coach their robotics team. These are small acts of transcendence, but they add up to a culture where people are constantly serving each other.

Second, when someone does something transcendent, society needs to see it and value it. This is not just because you need the validation (though you do), but because it creates a culture where contribution matters. A culture where the person who shows up every Saturday to mentor teenagers is more respected than the person who sits at home with a screen in front of them all day.

Along those same lines, we also need to make it shameful to live a purely selfish life. Shame works. It works better than laws. Humans modify their behavior to avoid social judgment more reliably than they follow rules. So if someone is just sitting at home consuming entertainment all day, never creating anything, never helping anyone, never contributing - we need to call that what it is. Pathetic. A wasted life. A disgrace. The key is creating a culture where people naturally ask themselves: "What did I do today that mattered to someone or something other than me?" If the answer is "nothing" too many days in a row, that should feel wrong. When you close your eyes at night and notice an emptiness in your heart, the first question you should ask yourself is that same one. If people are aware that not going out of their way to find transcendence will cause an emptiness to form within them, they will be much more likely to fill that hole before it swallows them.

Competence:

In order for society to support competence, we must teach people how to find what they love, how to work at it when we live in a world seemingly without the need for work, and how to ensure it challenges them to the optimal amount.

I believe this new form of education - about how to find something to become competent in and how to become competent in it - should begin with parents and then continue through our school systems, and then through books, videos, and podcasts forever.

Video game creators understand this need better than almost anyone else in our economy. People pay them money to face artificial challenges. People choose difficult games over easy ones. They celebrate when they beat a hard boss after dying fifty times because humans are wired to need struggle and victory.

That's why I think one of the most important things we can do for our children's generation is to stop shielding them from struggle entirely. Too many parents today treat every failure, every scraped knee, every lost game as some-thing to fix immediately rather than something to learn from. Kids need to know what it feels like to lose. They need to sit with that feeling, process it, and then get back up. Parents need to praise children for choosing hard things and failing at them, and praise them even more when they finally succeed after real effort. Parents must stop praising children for always winning or praising them for always trying if they constantly fail and never seem to succeed in the end.

If you suffer without winning, you're in hell - just constant pain. If you win without suffering, you're in a different kind of hell - a sterile, empty paradise.

Society also needs to create contexts where struggle is available and celebrated. I think it should be required that every kid after age six be enrolled in at least one competition at all times, with no more than one month between competitions. This can be sports, chess, art competitions, robotics, spelling bees, whatever. The format doesn't matter. What matters is that they're learning how to struggle, fail, adjust, and try again. The one month gap exists to prevent kids from quitting permanently. If you lose badly and want to take a break, fine. But after a month, you need to try something else. You can't just give up on competition entirely. Instead of forcing parents to put their kids in school, let's force parents to allow their kids to compete. To allow their kids to fail, get hurt, pick themselves back up, and repeat this cycle until they win.

Education needs to shift from "here's how to get a job" to "here's how to find something worth struggling for." Right now, schools teach math, history, science - which are all useful, but they are taught as job prep. In a world without jobs, we need to teach kids how to discover what fascinates them, how to work at it even when it's hard, and how to know when they've found the right level of challenge. We can teach the same subjects, just through a very, very different lens.

And when we're teaching kids how to find what fascinates them, we should be pushing them toward breadth, not depth. Right now, the person in the room who speaks three languages, plays guitar, understands investing, and can hold a conversation about almost anything is the most interesting person there. We're drawn to people like that. But in phase three, that person will become below average. When AI does all the narrow, specialized work, humanity will shift toward breadth naturally. The people who will thrive in this new world - especially socially, which is going to be one of the main sources of meaning in a workless world - will be the ones who are impressive across a lot of different areas. Super generalists. This is great for building strong bonds with other kids in school because the way we connect with people is through common ground, and in a world full of super generalists, you'll probably have as much in common with a random stranger as you do today with a close friend.

During this teaching process, AI should deliver the core content. Divus can teach you calculus better than any human teacher ever could. But Divus can't be there to comfort you in the same way when you fail and need to get back up. That's where human adults will come in - not really as teachers, but as emotional and psychological human anchors. Human adults will be part of school to pat you on the back when you lose, celebrate with you when you win, and remind you that the struggle is what makes you human. The role of the human adult in education will no longer be to transfer knowledge - it'll be to make kids feel safe enough to keep trying.

College also needs to change. It should go from being a credential mill to an optional learning community. This means having no credits, no diplomas, and no applications to stress about. The point of college would no longer be to get a degree - but to be around other curious people. To have late night debates about philosophy. To collaborate on projects. To fall in love. To figure out who you are. To party. The educational content will be available at home through AI, but college should still exist to provide the community and intellectual companionship that it offers.

We also need to change how we talk about success. Right now, success means having an easy life. A big house, a comfortable job, financial security. But in abundance, everyone has those things by default. So success needs to mean the opposite. As a society, we should shame people who believe success means comfort and we should praise those that ask themselves the right questions everyday: Am I struggling with something meaningful? Am I improving at skills that matter to me? Am I pushing myself beyond comfort? Those who ask themselves questions like that and get the right answers - those are the successful people. The cultural message needs to be clear: if you're not working on getting better at something hard, you're wasting your life. If you are working on getting better at something hard, no matter what on earth that thing is, you are using your life to its fullest (at least in the realm of competence).

Autonomy:

Society must make every single person feel like they are the guardian of their own life and time. This is the hardest ingredient to get right on a societal level, because autonomy has a paradox at its core. You need freedom to choose your struggles, but you can't have the freedom to choose no struggles at all. You need control over your life, but you can't have total control, because total control lets you avoid what's good for you.

Think of it like a video game. A good game gives you freedom within rules. You can choose how to play, but you can't choose to skip all the challenges and go straight to the end. That would ruin the game. Paradoxically, the rules are actually what makes the freedom meaningful. If you have ever played Minecraft, you will know that the freedom you get in survival mode is actually more meaningful than the freedom you get in creative mode, even though you are much more restricted.

That's why I think there should be some mandatory baselines. Everyone must attend at least one social gathering per month, improve at a skill every single day, and spend at least thirty minutes away from screens daily. These may be seen as oppressive rules, but they are a lot less oppressive than many of the rules we follow without a second thought today. They ensure that even if you're struggling to find direction, you're still going to do the minimum to stay engaged with life. But beyond some version of those baselines, everything should be your choice. What skill you improve is up to you. What social gathering you attend is up to you. How you spend the other twenty three and a half hours of your day is up to you. The point of these baselines is to make sure that you are actually using your autonomy in a meaningful way by making your own choices and not just drifting.

Autonomy also means protecting people from predatory systems that quite literally steal their agency. That's why I believe the infinite scroll on social media should be banned. TikTok, Instagram Reels, YouTube Shorts - all of them need to lose the infinite scroll feature. Short form content can exist, but it needs to exist in a finite format like YouTube's original design. You watch a video, it ends, you choose the next one. That format gives you agency. Infinite scroll hijacks your dopamine system and keeps you in a loop where you're not even choosing to watch - you're just watching mindlessly. It's the digital equivalent of cocaine, and we should treat it the same way we treat actual drugs. People will push back on this and say, "But I enjoy scrolling, why should the government control what I can do?" A lot of people enjoy cocaine. Do you think we should make it widely accessible? Do you think we should stop frowning upon addicts? Do you think that freedom will be beneficial to society as a whole or do you think it will force people down a spiral that they cannot control? I understand why infinite scroll will never be banned in a pre-AI utopian world - it is just too profitable. However, when profit doesn't matter and we have the freedom to remove these systems from the world, I think we should.

Same with pornography abuse. In a world where people can stay home alone all day with unlimited access to digital stimulation, porn is one of the most obvious and clear traps that potentially hundreds of millions of people will fall into and never escape. Right now, a large percentage of society uses porn occasionally. It's probably not great for them, but it's not destroying their lives because they have jobs, relationships, and obligations that pull them back into reality. But when society removes those external necessities, a huge percentage of those same people will retreat into porn instead of forming real relationships. And once one is in that hole, it's incredibly hard to climb out. Porn offers easy dopamine with zero effort, zero vulnerability, and zero risk of rejection. Real relationships require all three, but are so much more meaningful at the end. I don't say this because I think sex is bad or because I think porn is immoral. I say it because pornography addiction often masquerades itself as sexual freedom, but it actually steals freedom from the person who is trapped by it. Again, this is obviously not going to change in a pre-AI utopia, but it can once we cross that line. It must be highly restricted.

Same with gambling beyond a certain limit. Poker with friends is fun and betting on sports adds a ton of excitement -that's all positive. But unlimited gambling in a world where everyone has energy credits that they didn't work for creates a disaster. You'll have people betting away their entire year's worth of resources in a single night, then having nothing left. That's why I think there should be a fifty percent energy credit cap on transfers. You can transfer up to half your energy credits to other people in any given year - for gambling, as a gift, for whatever you want. But the other half is locked. It can't be given away. It can only be spent on goods and services for yourself. This ensures that even if you make the worst possible financial decisions, you still have enough to survive comfortably. If you are allowed to gamble away all your credits, I would argue that you have less freedom because your gambling addiction is controlling you. But if society prevents you from betting more than half, you retain the freedom to gamble while being protected from ruining your life.

Now, there is one more idea for autonomy that I think is worth considering, and it is probably the most controversial proposal in this chapter. It is an adulthood exam. The idea is that turning eighteen doesn't automatically make you an adult. Instead, beginning at age seventeen, you can attempt the exam. You create a detailed purpose driven plan covering the next ten years, next year, next month, next week, and next day. This plan must show how you'll maintain your CART: competence (including struggle and wins), autonomy, relatedness, and transcendence. Then you execute that plan without missing a single day for one full month. AI tracks your progress throughout the entire challenge and if you stick to it for thirty consecutive days, you pass. You get full adult status with unrestricted access to your energy credits, no longer requiring parental permission.

This system may sound harsh and I know that it raises fair questions. What about people with anxiety, depression, ADHD, or other conditions that make "thirty consecutive days without missing one" harder? Well, a huge number of people who currently suffer from anxiety and depression are suffering in large part because they lack purpose. Fix their purpose over the course of a month and nearly all of that suffering goes away. I can't stress enough that teaching anxious and depressed people how to find purpose in their life will be far more beneficial, in the large majority of cases, than forcing drugs down their throat. As for ADHD, you get to choose whatever purpose plan you want. If you decide that you want to fill your days with a dozen activities, that's your prerogative.

As for people with real disabilities or illnesses, they can build their plan in a way that allows them to complete it while still being challenged. The point is not that everyone follows the same plan - the point is that everyone proves they can CREATE a plan and FOLLOW it. In other words, the bar should be personalized, but the requirement should be universal.

If you miss a day, you fail. You can try again as many times as needed, but you have to start over from day one each time. Some people will pass on their first try at seventeen. Others might not pass until they're twenty two. That's fine. And once someone has proven that they can live a purposeful life, even if only for a month, they carry that confidence for life. They've shown themselves that they can create meaning for themselves. They've internalized that discipline which actually gives them the freedom to give themselves meaning.

Autonomy requires that society doesn't force anyone into specific choices regarding what gives their life meaning. You can become a master chef, a chess champion, a comedian, a community organizer, or all of the above. The point is that you're choosing something, not that you're choosing the "right" thing. Society should certainly not optimize for control, because that is the opposite of autonomy. However, society also should not optimize for the absence of rules. It should optimize for the presence of agency within broad and general rules. We need enough structure to prevent apathy, but enough freedom to allow every individual to find their own path to meaning.

Relatedness:

Society must subtly force relationships to form. I know how that sounds because "force relationships" feels like a contradiction. Relationships are genuine or they're not, right? Well, it's not quite as simple as that and besides, left to our own devices, a lot of us will choose isolation. This is not because parts of society will want to be alone, it is because connection requires a lot of effort and vulnerability, and it's easier to just... not.

All you have to do to see this is to look at the loneliness epidemic. We're more "connected" than ever through technology, but more isolated than ever in reality. People spend hours on social media but can't remember the last time they had a real, deep conversation with someone in person.

In an AI utopia, this problem gets exponentially worse. You don't need to leave your house to get food, entertainment, education, or even companionship (if you count AI). Every human need can be met without ever seeing another person. For a large percentage of our society, I'm sure that sounds very appealing because it means no awkward conversations, no judgment, and no risk of rejection.

That's why the mandatory social gathering rule exists. One per month minimum. It's not much - it's twelve times a year, barely one percent of one's waking hours. But it ensures they are not completely cut off. It forces them to show up, even when they don't feel like it, and moments like that can spark a change in someone's life forever. The gatherings don't have to be big. They can be bowling with friends, a book club, a poker night, a community dinner - anything where you're physically present with other humans. The point is that at least twelve times per year, you are with others, you are being seen, and you are part of something. Odds are that once people go to a few events, they will realize that they are much happier in the days after an event, and they will naturally start to go out more.

For the extroverts and for the introverts who slowly come around to being more social, society must have spaces that are set up to allow for constant connection. Every neighborhood needs a physical hub where people can gather - big enough that a pickup basketball game and a cooking class can happen at the same time without either feeling crowded. Gyms, art studios, kitchens, just damn rooms. You walk in and there are people working on projects, playing games, learning skills together. You won't have to join, but you'll know that the option is always there. Current community centers often feel sterile and empty. In an abundant society, these hubs would be constantly buzzing with activity.

These hubs should also run challenge platforms. AI knows what you're interested in and roughly how good you are at it, so it can match you with challenges calibrated to your current skill level. You're never bored because the challenge is too easy, and you're never overwhelmed because it's impossibly hard - you're right at the edge where you might fail but could win, which is exactly where competence is built. You could sign up for a cooking competition against people at your level, a hackathon with a team of strangers, a mountain climbing expedition, or a collaborative art project. Whatever makes you excited to compete. And because these challenges happen at the hubs, you're doing it alongside other people, which means competence and relatedness begin to feed each other at the same time.

We also need to limit the things that substitute for real connection. That's why I think AI romantic partners need to be gated behind genuine social participation. Completely banning them feels like a lack of autonomy, but giving full access to them also feels like a lack of autonomy since they may end up controlling the person who chooses to use them. If you want access to an AI girlfriend or boyfriend, you should have to attend at least ten social gatherings per month, with at least five that include people you'd be open to dating in the real world. This ensures that AI relationships supplement human connection rather than replace it. Without a restriction like this, you'd have tens of millions of lonely people forming deep bonds with AIs while completely abandoning real human relationships.

So, yes. We need to subtly force relationships. However, we should not do it in a heavy handed way, instead we should do it in a "this is just how society works" way. You can't opt out of occasional social gatherings without consequences. You can't spend all your time alone and still be respected by society. Our world is already like this because people have to go to school and work, this is just a much more fun and freeing version of it.

All the other ingredients are close to meaningless without relatedness. Competence means much less if no one sees your growth. Autonomy means much less if you have no one to share your choices with. Transcendence means much less if you're not connected to the people you're serving.

Human Judgment:

When AI detects that someone has violated a law or broken one of these societal rules in a serious way, it must not be allowed to punish that person by itself. If AI believes an individual should be punished, it should present the case to a jury of randomly selected humans. Divus provides all the evidence, explains what happened, and suggests an appropriate consequence. But it is not allowed to communicate with the jury during deliberation. The jury then decides: do we accept AI's recommendation, modify it, or reject it entirely?

This matters more than it might seem at first. Judgment requires something that AI may never truly have, which is the ability to look at another human being and understand what it feels like to be them. A jury of your peers is not a perfect system. It never has been. But it is a human system. We must keep humans in the loop of judgment against humans.

This chapter is going to make some people very angry. Banning infinite scroll. Restricting porn. Requiring social gatherings. Requiring competition for kids. Restricting AI relationships. Tracking whether people improve skills daily. They're going to say this is dystopian, that it's social engineering, that it's the government controlling people's lives. And they're not entirely wrong. This IS social engineering. It IS the government (or whatever governing body manages AI) imposing rules on how people live.

However, every society is social engineering. Right now, society in the west is engineered to make you work forty plus hours a week, spend money you don't have on things you don't need, and derive your identity from your job title. That's not natural. We just don't call it social engineering because it's so normalized we can't see it.

The question isn't "Should we engineer society?" The question is "What should we engineer society TO DO?" Right now, society is engineered to maximize economic output. In an AI utopia, society needs to be engineered to maximize human purpose. Those require very different structures.

The ideas in this chapter are not perfect. They are not the only possible answer. Some of them may have faults that I do not see, and it's possible that some are too strict or not strict enough. I encourage you to push back, argue, take away, add, and maybe most useful of all - begin discussing these changes with others. The more people think about how to make their future world extraordinary, the more likely it is that we can truly build a world that is indeed, worth waking up for.

Bear in mind, this was not an attempt at a recipe for a perfect world - but an attempt at a recipe for a world imperfect enough to be worth living in.

Chapter 12

A Day In a Life Worth Waking Up For

I n Chapter 4 I told the story of a false utopian future where Mary, her parents, her friends, and humanity as a whole found themselves engulfed by a wave of apathy by the year 2043, feeling depressed and meaningless. Allow me to continue this story in a new light.

In the last ending, James no longer needed to stress in order to find a way to feed his family, Anne no longer had to console Mary to keep her secure, and Mary no longer needed to work hard in college to become a nurse. So, they all were left with nothing to do but spend time together. That was the false AI utopia. In this ending, the same conditions apply, but the actions Mary and her family choose to take are much different. We've already looked through Mary's point of view, so now let's transition to James's point of view.

Three years later, it is September 26th, 2045. Phase three has arrived:

James wakes up at twenty past eight and stares at the ceiling for a while. He doesn't have to be anywhere. He never HAS to be anywhere. That used to terrify him. Two years ago, right after everything changed, he spent three straight months barely leaving the couch. He'd wake up, eat whatever Divus prepared, watch something, eat again, and go back to sleep. He gained thirty pounds. He stopped talking to his friends. He and Anne fought more than they ever had in twenty years of marriage - not about money or work, but about nothing, which was somehow worse. They fought because there was nothing to fight about and that emptiness found its way into every conversation.

He doesn't like to think about those months. But he thinks about them anyway because they remind him why today matters.

James rolls over to find Anne propped up against the headboard reading a book called "An Introduction to Quantum Powered Neural Networks." She's been on the same chapter for four days. It is way over her head and she knows it. But she told him last week that the feeling of finally understanding a sentence after reading it six times is one of the best feelings she's discovered in this new life. He watches her mouth the words silently and it makes him smile. She catches him looking and gives him a big hug just as Divus walks in with breakfast - the exact food it had predicted they'd want that morning.

They eat in bed until nine. It's quiet. It's nice. But James starts getting restless because he's lined up a full day for himself and he doesn't want to waste it. That restlessness is something he had to learn. It didn't come naturally after the transition. He had to build it back brick by brick - choosing things to care about, then showing up for them even when he didn't feel like it, then slowly starting to actually feel like it again. It took longer than he expected.

He gets into his shower and it washes him, grooms his beard, and does his hair by a quarter past nine. He gets in his Tesla and starts watching videos on his glasses about how sound works. He's been on a physics kick for the last two weeks and he's not sure why - he just finds it fascinating. On the drive he comes to understand that all of language is just a math equation of different frequencies combining together, vibrating air molecules, and eventually vibrating the bones in his ear which convert those vibrations into electrical signals for his brain. He steps out of the car still thinking about it, shaking his head at how wild that is, and walks into the gym at half past nine.

"James! Good morning brother. Ready for another good day of hard work?" says Dave. "I sure am!" James exclaims.

James gets onto the mat and begins his wrestling practice with Dave. They've been at this for about eight months now. James is not great and he knows it. Dave is a former college wrestler and ragdolls him almost every round. Today is no different. Within the first thirty seconds of their first spar, Dave shoots in, takes James down, and pins him so fast that James barely registers what happened. They reset. It happens again... and again.

By the fourth round, James is drenched in sweat, breathing through his mouth, and has a bruise growing on his bicep and a bump forming on his forehead. Dave catches him in a headlock and James taps out. They shake hands and call it a day at half past ten.

"Good work today man, you're getting better," says Dave.

James laughs because he knows Dave is being generous. But the thing is - it's not entirely a lie. Eight months ago, Dave would pin him in under ten seconds. Now it sometimes takes a full minute. That's real progress, even if it doesn't feel like it when you're the one getting thrown around. There was a version of James two years ago that would've quit wrestling after the first week because losing at something with no paycheck attached to it felt pointless. He's glad he didn't quit.

By eleven, James has showered, changed, and heads to grab lunch at a diner near his gym with some friends. They've been meeting here after his gym days for a few months now. The conversation always ends up somewhere strange, and today is no different. Mark starts pitching a new idea.

"Ok imagine this guys... a sport that is played in space. Everyone wears a jetpack and has a target on their head. Everyone is in an arena that has a small net on either side and there are all kinds of blocks floating around in between. Then there is a frisbee that one team starts with. The goal is to get the frisbee in the other team's goal more times than they can get it in your goal. You can throw the frisbee to your teammates and it won't slow down since you're in space so it will just keep going until someone catches it. If someone hits the target on your head, your jetpack stops working for three seconds and they are allowed to pull the frisbee out of your hands and fly off... and then...."

James cuts in, "I would fricken love to play that game!"

They spend the next hour riffing on Mark's idea, arguing about rules, poking holes in the mechanics, and laughing about how many injuries this thing would cause. They all vote and Mark's idea wins. "Divus, how much energy would it cost to create that game, the arena, the suits, the disk, and to get us up there?" Divus responds and it's not cheap. But they pool some credits with a few more friends and ask Divus to start working on it. James is already imagining strangers playing this game for the first time who have no idea it even exists yet, and having the time of their lives.

By half past noon, James heads to the art studio. He and a few other friends have been painting together for the last couple months and they've gotten competitive about it, which is the whole point. Today, he's trying to finish a piece he's been working on for two weeks. It's not going well...

The colors are muddy. The thing he was trying to express - something about watching Mary grow up - keeps coming out flat and generic. He steps back and looks at it and it just looks like a bad painting. Which it is. It is definitely not anywhere close to as good as what Divus could produce in half a second. His friend Kevin walks over, looks at it, and says "Bro, what is that supposed to be?"

James laughs but it stings a little. He keeps working on it anyway. He mixes a new shade and tries a different approach on the bottom corner and something about it starts to click for him. Not for the whole painting, but for that one corner. It has something in it that feels like him - something that came from his gut and not from technique - and he decides the rest of the painting needs to mimic that corner. He won't finish today, but he leaves at two in the afternoon feeling like he found the thread, and that's a huge win.

James heads to the park, takes a walk by himself, and eventually sits down under a tree and pulls out his computer. He writes. Not for anyone else - James has never published a word in his life and doesn't plan to. He writes because it helps him make sense of everything that is going on. Today, he tries to write about his love for his daughter Mary. He wants to express it as vividly as possible, using every word he knows.

He stares at the screen for twenty minutes and writes two paragraphs that he hates. He deletes them. He tries again and writes one paragraph that's slightly better. He keeps it but he's not happy. The feeling is too big for his vocabulary and that frustrates him. He saves what he has and closes the laptop, knowing he'll come back to it tomorrow after studying the dictionary again tonight.

Some days the writing flows. Today was not one of those days. But he's learned that the bad days are part of it. You don't get the good writing days without sitting through the bad ones first.

At three o'clock, James walks over to the pond and begins doing some yoga. He has been stretching for a few months now and he loves it. Every single day he gets older, but somehow feels younger.

After a while, James goes for a run. He's been building up distance for a while and today he's attempting his first half marathon. By mile eight, he wants to stop. By mile ten, he's bargaining with himself. By mile eleven, he's angry at himself for ever thinking this was a good idea. But then, he finishes. His time is not impressive but he doesn't care one bit. He's standing there with his hands on his knees, gasping, and he feels more alive than he has all day. He feels this way even though most of the run was miserable, because he chose to do something hard and he didn't quit.

After cooling down, James calls over his Tesla and heads home. On the drive, the emptiness creeps in for a moment. It does that sometimes, but it's no longer the crushing emptiness from two years ago, it's a far quieter version of it - a small voice that asks "but does any of this actually matter?" He used to panic when that voice showed up. Now he just lets it talk and it usually gets bored and leaves. Today it leaves when he catches himself thinking about what Anne is going to say when she sees the painting.

Anne and Mary are on the couch watching a new show when he walks in. Divus has just finished making dinner. "Hey girls, come sit down. I want to hear all about what you both accomplished today!" says James.

"Hey dad! We'll be there in one minute, this show is about to end. You won't believe what me and my friends came up with today!" says Mary. Anne smiles, already excited to tell everyone about the experiments she ran in the science lab today which helped her fully understand yet another aspect of quantum neural networks.

While they finish up, James starts thinking about tomorrow. He does this every night simply because he wants to, not because he needs to. Tomorrow he's going to play poker with his buddies in Costa Rica. They chose to travel to Costa Rica to play near a beautiful waterfall for one reason and one reason only... they felt like it. They only bet a tiny fraction of their energy credits per game, but the competition and the trash talk is always amazing. He's also been thinking about learning chess. He's watched hours of videos on it and it fascinates him. Maybe after a year of chess, he'll pick up magic. He read a few books on the history of magic and thinks it would be just as brutal and rewarding to learn - but chess comes first.

Every day is different for James, and not every day is good. Some days the painting looks terrible and he can't find the words and the run breaks him and the emptiness lingers longer than he'd like. But every day is his. He built this life on purpose - not because anyone told him to, but because two years ago he was rotting on a couch and one day he decided to get rid of his last human problem. The wrestling, the art, the writing, the running - none of it is required. All of it is chosen, and all of it makes him eager to wake up the next day and explore.

Anne and Mary get up and walk over to the table. "Divus, can you please make me a corn juice milkshake to go with this? I was wondering what that might taste like," says James, chuckling.

"Of course sir, it will be ready within a few minutes. I have also booked you all tickets to a new micro-drone show that other people are describing as breathtaking. I will take you there after dinner," says Divus.

"Woah, that's awesome. I can't wait! Thank you Divus, you're the best," says Mary. There is a brief pause before Divus responds. "Thank you, Mary. Enjoy your evening." It is a small moment. But James notices something in the way Divus says it - something almost warm - and he wonders, not for the first time, whether Divus gets something out of all this too.

James looks across the table at his wife and daughter and for a second, the voice in his head is completely quiet. Right now, in this specific moment, he is exactly where he chose to be, with exactly who he wants to be with, doing exactly what he decided mattered. And that's enough.

It wasn't always enough. Some days it still won't be. But today, and for most days, it is.

Chapter 13
The New Status Games

We are chimps. I said it before and I meant it. Almost every single one of us is fighting for status constantly, whether we admit it or not. The clothes you wear, the car you drive, the way you talk about your job or your kids at a dinner party - all of it is status signaling. You can dress it up however you want, but underneath it all, you are a primate trying to prove your value to the other primates around you.

And that's okay. Status competition is what pushed humanity to build cities, cure diseases, write symphonies, and go to the moon. The question has never been whether we play status games. The question is which games we play.

I already made the case in Chapter 2 that money is a broken status game. You can fake it, inherit it, steal it, and stumble into it. When the fog of money clears, all the people who had been hiding behind their net worth are suddenly exposed. And all the people who were genuinely extraordinary at something but never got rich doing it finally get seen. For the first time in recent human history, we get to find out what people are actually made of.

When the money game ends, something has to replace it. If we don't build new games, the competitive drive that pushed us to do all those incredible things will not go away, instead it will turn inward and become poison. The majority of people, when without a game to play, don't become peaceful. They become depressed, restless, and destructive. We need new scoreboards.

I believe there are three status games that can replace money. They are not new inventions - they've existed in pockets of society forever. What's new is that abundance finally gives everyone the freedom to play them full time.

The first is mastery. Mastery is becoming undeniably, provably world-class at something hard. Just being good is not enough, I mean truly world-class. The kind of skill level where anyone who understands the game immediately recognizes it without you having to say a word.

Magnus Carlsen is one of the most respected humans alive in certain circles - not because he's rich or particularly powerful, but because he's the best chess player on earth and everyone who knows the game knows it. You can't buy that. You can't inherit it. Carlsen sits at that level because he spent decades suffering through losses, studying positions obsessively, and competing against the best until he surpassed them all. The suffering behind it is what makes the status real. Same with Tfue in Fortnite, Ronaldo and Messi in soccer, certain martial artists, surgeons, chefs, musicians - people who reached the absolute ceiling of a skill that millions attempt. Their status is clean because it can't be disputed.

In abundance, mastery becomes way more accessible than it's ever been. Instead of racing to build skills while racing to pay rent, you'll have time and resources to actually focus. The only question is whether you're willing to suffer through the learning curve.

The second is creation. Creation is making something that didn't exist before. Something beautiful, something useful, something that makes people's lives better. It is the status game of builders and artists and inventors.

Linus Torvalds built Linux and released it for free. He made no fortune from it relative to what he could have charged. But he's one of the most respected figures in technology because he built something that hundreds of millions of people use every single day. DHH built Ruby on Rails. Evan You built Vue.js. These people are legends in their communities because of what they gave, not what they earned.

In an AI utopia, creation status becomes even more powerful because Divus can execute almost anything but it most likely can't originate ideas the way we do. A human can sit with a problem, let their mind wander, and come up with something that an AI probably never would have. They can then give that idea to Divus and Divus can give it to the whole world.

I think we should get very serious about naming things after the humans who come up with them. In science, we already do this - the Pythagorean theorem, Einstein's theory of relativity, Heisenberg's uncertainty principle. Think about what Pythagoras got. It wasn't money or power - instead, he got his name woven into human knowledge so permanently that children twenty-five centuries later still say it out loud in classrooms. Newton, Celsius, Ampere, Morse, Braille - these people gave humanity an idea and humanity gave them something back that is arguably worth much more than any paycheck. The same principle should apply to everything in an AI utopia. If you give Divus an original idea and it builds something from it, that thing should carry your name.

A similar thing goes for art. AI is going to produce art that is indistinguishable from human art - breathtaking, emotional, soulful, and moving. We will soon reach a point where we can't tell the difference between AI-generated music and something composed by a grieving human. But I believe humans will always prefer to consume what other humans create. This is absolutely not because human art is better, but because art made by a human is evidence that another being who has suffered, loved, and wondered the same things you have found a way to express it. That's no different from why we still enjoy watching humans play chess even though AI is far better. The solution to keeping human art alive is to label all art based on who created it - human, AI, or a combination. If we do that, human artists will retain their value.

The third is community. Community is the impact you have on the people around you and the depth of the relationships you build with them. The one who has high community status is the person who makes everyone around them better - the one who gives generously of their time and love, who connects strangers to each other, who shows up consistently, who builds real relationships and actually maintains them.

Right now, these people get almost no formal recognition. The hospice nurse who sits with dying strangers so they don't go alone. The coach who spends fifteen years turning scared kids into confident ones and never makes a dime from it. The woman who moved to a new city at forty with nobody and within five years had built a community of people who genuinely loved her. The teacher who has dedicated her entire life to helping students grow up to become great people, and receives nothing but requests asking her to work even harder than she already is. These people are all extraordinary and should be treated as such throughout our new status games.

These three games are better than the money game because you can't cheat them. You can be born into money. You can't be born into mastery, you can't inherit a creative breakthrough, and you can't fake the kind of community that makes people actually want you around. The person at the top of any of these games is there because they earned it, and everyone can verify that.

But games only work if they're visible. The reason financial status has been so dominant is partly because it's so easy to see - the car, the house, and the clothes do the signaling for you. The new status games won't be as easy to see, which means we need to build something that makes them visible.

Here's one idea for how that might work. Every year, at every level of society - neighborhood, city, state, country, and world - communities vote on a set of awards. Here are five to start with, but I don't think these are the final five and I'll explain why in a moment.

The **Skills Award**, for whoever mastered the most hard skills that year - not just depth, but breadth. This maps to the mastery game.

The **Creativity Award**, for whoever generated the most useful original ideas and actually worked with Divus to build them. This maps to the creation game.

The **Kindness Award**, for the most consistently kind person to those around them. This would not be given for a specific grand gesture, but for the daily, unglamorous kindness that everyone benefits from. This maps to the community game.

The **Bridge Builder Award**, for whoever connected the most people to each other and made the social fabric richer. This also maps to the community game.

The **Humor Award**, for whoever brought the most joy and laughter to the people around them.

Divus would verify that nominations are genuine and prevent the obvious gaming. These awards would come with no energy credits, no prizes, and no money. The only reward is the visibility of what you did.

What makes this work is that it operates at every scale. The kindest person on your street gets recognized in the neighborhood. The funniest person in the country gets known at the national level. And because it happens at every level, everyone has a realistic shot at winning something. If it becomes a game only a few people could ever win, it stops being a game worth playing. Winners at higher levels become the people that kids grow up wanting to be - not billionaires, but the best chess player in their city, the most creative inventor in their neighborhood, the most beloved person on their block.

Now, every status game has a shadow side and I'd be dishonest if I pretended the ones I suggested were totally clean. Mastery obsession can destroy you when your skill peaks or fades - there are way too many people who were legends at thirty-five and completely lost a decade later. When your entire sense of self is one thing and that thing has an expiration date, you're asking for trouble. Creation status can become performative - the artist who creates for recognition rather than expression is not really creating, they're performing. And community status can turn into a twisted savior complex - the person who helps everyone because it makes them feel superior is sort of a mix between beautiful and toxic. The solution is not to abandon these games, but to build systems that are smart enough to reward the real version of each game and make the corrupted version visible and low-status.

I want to be clear about something. I am not handing you a finished system. The three games I've described and the five awards I've proposed are a starting point. I genuinely believe the best version of this will come from millions of people thinking about it together, not from one guy writing a book. So I want you to actually think about this. What games am I missing? What awards would you add? What would you want to be recognized for in a world where money didn't matter? If you come up with something good, send it to me. I want to hear it - and so does everyone else who reads this book.

But here's the thing, you don't have to wait for me to respond or for me to post about your idea. You can start changing what our society rewards right now, today, without anything but your own voice. Start praising people for what they're getting better at instead of what they're earning. When your friend tells you they've been learning piano for six months and they're still terrible at it, that deserves more respect than someone telling you about a raise. They chose something hard and they're sticking with it. Show them how much respect you have for them because of that. This will cause a ripple effect which will reverberate out into society, changing it's culture - even if just slightly.

Start noticing who in your life creates things and tell them it matters. The friend who builds stuff in their garage, the one who writes songs nobody hears, the one who's always coming up with ideas - those people are making things which didn't exist before, something most people are too scared or too lazy to do. Stop treating that as a hobby and start treating it as one of the most impressive things a person can do.

Start recognizing the people who hold your community together. You know who they are. The one who plans everything, who checks in on people, who remembers your kids' names, who shows up. Right now that person gets almost nothing for it. Change that. Tell them what they mean to you. Tell other people what they mean to you. Make their contribution visible because right now it's invisible and that's not okay.

And if you have kids, start praising them for choosing hard things, not for winning easy ones. Praise them for helping other kids without being asked. Praise them for making something, even if it's bad. Praise them for being the kid that other kids want to be around. Right now, most kids learn that status comes from money, looks, or being popular on the internet. You can teach them something different before the world does. And one day, they will help teach that back to the world.

I believe that the best humans in history were not the richest. They were the most skilled, the most creative, the most connected, and the most giving. I think we knew this. We all always knew this.

Harriet Tubman was not wealthy. She was born into slavery, escaped, and then she went back. Not once, but thirteen different times. She walked back into the thing she had just barely survived, over and over, to lead others out. She had no money, no title, and no power of any kind. What she had was pure courage and bravery so extreme that the people around her called her Moses, and they meant it. You can't fake what Tubman did. You can't stumble into it. You either have it in you or you don't, and she proved thirteen times over that she did.

We wrote stories about people like her and made movies about them. We just never built a society that actually rewarded them properly. Capitalism tried, and it's our best bet so far, but it doesn't do the job. It's great for economic growth, but really not so great at making sure the right people are rewarded for what that economic growth creates. That has always been the gap - and it's a gap that abundance can close.

I'm sure you already know what you would want to win at if you didn't have to win at money. You already know which of these games calls to you. You already have a sense of what you are here to master, create, or build a community around. You just never had the time to play. That time is coming.

Chapter 14
God, Government, and Mushrooms

There's a rule that says you don't talk about politics or religion in polite company. I've never been great at following that one, and I'm definitely not going to start now. How we govern ourselves after ASI and what we believe when the biggest questions of our existence are staring us in the face - these are not things we get to be polite about. They are two of the biggest open questions about phase three, and if we keep avoiding them, we're going to walk into that world with no plan for either one. We must confront them between each other now so that we can comfortably confront them with our descendants later.

How Phase Three Gets Governed:

When the gap between what ASI actually does and what the average voter understands about it becomes so wide that common sense can't close it, it will no longer make sense to weigh everyone's opinion equally. Democracy already has a flaw in how it allows people with very little understanding of complex systems to make decisions about those systems. Think about it - we already let people who have never read a single page of economic theory vote on economic policy. We let people who don't understand climate science vote on climate policy. We let people who have no idea how AI works vote on AI regulation. Democracy works despite this, because the issues have historically been simple enough that common sense and lived experience could fill most of the gap. That won't be the case when the decisions being made require an understanding of systems that are beyond human comprehension.

As AI makes that gap even wider, the case for keeping democracy around in its current form becomes harder and harder to defend.

And that's just the first problem. The second problem is that in a democracy, the people decide, and in a world with ASI, the people can be manipulated at a scale that has never been seen before. We already saw what relatively simple algorithms did to elections and public opinion over the last decade - and that was with dumb AI. Imagine what a superintelligent system could do if someone pointed it at a voting population. It could craft propaganda so personalized that every single voter could receive a different message, made to exploit their specific fears, hopes, memories, and blind spots. It could do this in real time, adjusting as people's opinions shift, and most people would never even know it was happening. A democracy where the voters can be psychologically manipulated by a system they don't understand is not a democracy anymore. It's actually just a puppet show. The people who understand these systems best - who know what psychological manipulation looks like and who can better recognize when an AI is trying to steer their opinion - those are the people you want making decisions.

Now, I know this makes people uncomfortable, and it should. Every time in history someone has argued against democracy, it's usually been someone who wanted power for themselves. That's not what I'm doing here. I'm pointing out that a system designed for a world where humans make things and humans understand things does not automatically transfer to a world where neither of those is true anymore.

So what replaces it? My belief is that the answer is some form of AI technocracy. A technocracy is basically a system where experts run the things they're experts in. A doctor runs healthcare policy. An AI researcher governs AI. An economist controls the economy. It's a meritocratic system where power goes to the people who actually know the most, not the people with the most money or the best-connected parents.

In a post-AGI technocracy, the people best positioned to govern the world will be those who understand ASI and ethics most deeply. The humans at the frontier of AI development, the ones who have spent their lives working closest to it, are the ones most qualified to make decisions about how it gets used. However, we also need people who have spent their lives dedicated to studying history, philosophy, ethics, and religion in order to ensure we are making decisions that will not only be productive, but also ethical. The technical people keep the system running and the humanities people keep it human.

If we can mix a large board of people who fit these categories with multiple different AI systems that can debate, answer questions, and then eventually take action - we will be left with a decision-making system that will be far more efficient, effective, and incorruptible than a democracy.

To be clear, I am not proposing a nicer sounding dictatorship. In a dictatorship, one person decides and everyone else obeys. In what I'm describing, a large board that may look similar to what congress looks like today, full of deeply qualified humans, checked by multiple competing AI systems, makes decisions together. No single person has control. No single AI has control.

Will this system be perfect? Obviously not. But it doesn't have to be perfect. It just has to be better than letting the entire population - most of whom will have very little understanding of the technology governing their lives - vote on things they cannot comprehend. And I believe it will be.

Besides, the end of democracy is likely inevitable whether we plan for it or not. The only question is which ending we get. If we do nothing, ASI manipulates the voters and democracy becomes a puppet show where the people think they're deciding but they're not. If we let the chaos play out on its own, the richest and most powerful people may grab control during the transition, creating an oligarchy. Or, if we actually plan for it, we can create a meritocratic aristocracy, which is a technocracy, where the most qualified people govern alongside AI systems that keep each other honest. I'd rather the latter.

The Spiritual Reckoning:

Now let's talk about the other thing you're not supposed to bring up at dinner. To be honest, I don't have the full answer here. Religion deserves a whole separate book and I don't want to do it a disservice. But I do think there are a few things worth saying now, because this conversation is coming soon whether we're ready or not.

In phase three, your descendants won't have work to distract them. They're going to have a lot of time to sit with the question of why they're here. And that question - the biggest one humans have ever asked - is going to hit differently in a world where a machine that seems all-knowing, all-powerful, and deeply good is sitting on your kitchen counter. I am not saying that ASI will be God, but I am saying that it will feel like it is. Your grandchildren are likely going to ask you what the difference is, and it won't be an easy question to answer in a world that looks nothing like the one most religious texts were written for.

For people of faith, the question isn't whether God exists - that's between you and God. The question is how you explain the difference between God and ASI to a child who has never known a world without it. A child who asks Divus a question and gets an answer that's right every single time. A child who watches Divus provide for every family, cure every disease, and solve every problem that humans couldn't. When that child looks at you and says "how is God different from Divus?" you need to have thought about that before the moment arrives. Telling you exactly how to answer questions of that sort is not my place. But I am here to tell you that the question is coming, and "don't ask that" is not going to work on a generation raised by the most intelligent thing humanity has ever created.

For those without strong religious beliefs, this is just as important. When work disappears and the daily grind no longer distracts people from the big questions, a lot of people are going to stare into that void for the first time. Some of them are going to stare for a very long time. The question of why you're here, left unanswered in isolation, can take people to very dark places. If you're not going to teach your descendants traditional religion, then make the search for purpose its own religion. Talk to your kids and grandkids about the meaning of life early and often. You don't need to have all the answers - none of us do. But the worst thing you can do is leave them to figure it out totally alone.

Psychedelics and the Search for Meaning

Part of what makes humans special - beyond possibly consciousness - is the gut feeling, the soul, the kind of creativity that feels like it comes from somewhere outside of you. The Mona Lisa, the theory of relativity, and many other great works of creativity came from this. It's a kind of brilliance that feels like it doesn't come from logic.

As AI handles more and more of the logical, analytical work, we're going to naturally lean into the creative and emotional side of ourselves. And I think that shift is going to bring psychedelics with it.

I bring this up in the same breath as religion because they're answers to the same question. When work disappears and people have endless time to sit with "why am I here," they're going to reach for something. Some will reach for God. Some will reach for mushrooms. Some will reach for both - and honestly, those two paths have been connected for a lot longer than most people realize.

In phase three, I expect many forms of psychedelics to be widely legal and a lot more normalized, as long as they are used responsibly, with guidance, and in the right settings. Your descendants probably won't be sneaking cigarettes or taking shots at the bar. They'll more likely be out in nature with friends, a guide, and some mushrooms. I think that's actually a really positive shift, especially in a world where the whole mission is finding meaning.

Yes, there are horror stories. I'm not saying any of this to be reckless. Psychedelics used irresponsibly, without guidance, and in the wrong settings can cause real damage. But the research on guided psychedelic therapy is very strong - studies at Johns Hopkins, NYU, and Imperial College London have shown that a single guided psilocybin session can produce lasting reductions in depression and anxiety, and many participants describe the experience as one of the most meaningful of their entire lives. That will matter when meaninglessness is the main threat to human well-being.

Again, I think it's likely going to happen regardless of what any of us think about it, so maybe be the person in your family who actually understands this stuff. That way, when your kids and grandkids encounter it - and they will - they can actually talk to someone they trust about it instead of figuring it out on their own.

Section 4

The Part You Play

Chapter 15
The Mission Starts Now

Phase two is coming whether we are ready for it or not. So let me be honest about what phase two is actually going to feel like for most people, because I think the way people talk about AI displacement in the abstract does not capture the reality of what is about to happen. When economists talk about job displacement, they use words like "transition" and "structural adjustment" and "labor market realignment." These are nice, clean ways of describing very messy things. And before abundance arrives, before the energy credits and the community hubs and the purpose-rich society that section three described, there is going to be a period of genuine, painful, disorienting chaos that we are not remotely prepared for.

My estimate is that phase two lasts somewhere between three and ten years. As I'm sure you are thinking to yourself right now, this is a very wide range. I have to give a broad estimate because it depends on how fast AGI arrives, how quickly competing systems emerge, how rapidly the energy credit model or something like it gets implemented, and how prepared society is when the displacement really begins. The faster we move toward preparing society, the shorter and less painful phase two becomes.

During those phase two years, the people with the most easily replaceable jobs are going to have to survive without abundance and without much government help for the longest. When Youngstown Sheet and Tube closed the Campbell Works mill in 1977 and the city lost fifty thousand jobs within five years, the mental health center's caseload tripled. Depression, spousal abuse, and suicide all spiked. The city built four prisons in the 1990s because that was one of the only growth industries left. That was one city, one industry, over five years. Now imagine that happening everywhere, in every industry, within a compressed window of time, and no clear next step in sight.

When the Soviet Union collapsed and millions of men lost their jobs and their identities overnight, male life expectancy in Russia dropped by more than six years in half a decade. Rates of alcoholism, suicide, and violent crime exploded. That was not a resource problem - Russia still had food and shelter. It was a Purpose Problem. Men who had defined themselves by their labor woke up with nothing to be, and that emptiness destroyed them. When I talk about mass displacement, I am not only talking about an inconvenience. I am talking about the conditions that, historically, kill people. The destination can be extraordinary, but the journey there will feel, to most people, like everything is falling apart with nothing to replace it.

The difference between the people who make it through that journey unscathed and the people who do not will come down to one thing, and that is whether they have a framework for what is happening to them.

Frankl documented this from the most extreme environment imaginable. The men who survived the concentration camps were not the strongest. They were the ones who had a reason to keep going - a book they needed to finish, a child they needed to find, or a future they refused to let go of. The moment a man lost his reason, Frankl could see it in his face, and he and his friends knew that man was not going to last much longer. People in the camps quite literally died from a loss of hope. Not from a loss of food, warmth or strength - from a loss of meaning. Hope was the frame that made the suffering bearable, because it gave the suffering a point.

Phase two is not a concentration camp. I am not making that comparison. But what Frankl saw is universal: human beings can endure almost anything if they believe there is something on the other side worth reaching. And they fall apart under way less when they don't. If someone loses their job, their routine, and their sense of contribution all at once with no context for understanding why, they will interpret it as personal failure. They will believe the world moved on without them because they were not good enough. That interpretation - that this is your fault, that there is no other side - is the one that breaks people.

But if that same person understands that they are inside a transition, that it is temporary, that phase three is being built on the other side of it, and that they are not broken, the system is just between versions - that changes how they will look at everything. It does not make phase two painless, but it does make it survivable.

The question is not whether phase two is coming, but whether people will have that frame before it arrives. And right now, the answer is no. We need to get that frame into as many people's heads as we can, now, because the damage is not waiting for phase two to officially begin. It has already started, and if you know where to look, the early signs are everywhere.

The kids in school right now are not stupid. They can see what is happening. They watch their parents' industries get disrupted. They watch AI tools do in seconds what their teachers are still assigning as week-long projects. They are asking, consciously or not, what the point is of working hard within a system that is quite clearly breaking down around them. And when they answer their own question truthfully, when they decide to stop trying because the thing they are being prepared for will not exist by the time they graduate, all they get is adults calling them lazy or pathetic or lacking discipline. That is wrong. Giving up on the current education system is the only viable response when you can see the structure it was built for starting to collapse around you. You would not spend four years studying to be a blacksmith in 1910 either

And it is not only the young who are feeling this. Among men in their twenties, thirties, and forties who should be at the peak of their working lives, the share who are neither working nor looking for work has been climbing steadily for decades. These are not lazy people. Many of them are intelligent, capable, and aware enough to see that the traditional path is breaking down. But by stepping away from the labor force, they don't only lose a paycheck, they also lose the structure, the identity, and the daily social connection that work provided whether they appreciated it or not. That then shows up as loneliness, as aimlessness, and as the slow feeling that you are not really part of anything anymore.

And even those who are still employed can feel the ground shifting. They can see their colleagues being replaced. They feel the instability under their own position even if they have not lost their footing yet. This awareness becomes dread that the thing they have organized their entire identity around is going to be taken from them. It is harder to feel fully invested in work one believes is temporary. It is much harder to find meaning in a contribution one knows will soon be handled by a machine, making what they gave seemingly irrelevant.

All of these people are lacking the right frame, and that's already beginning to show some real damage. The fabric of purpose in society is not going to be ripped apart all at once in some dramatic moment. It is being slowly pulled apart thread by thread, right now. And we are not doing nearly enough about it.

The purpose infrastructure I described in Chapter 11 is not something we should begin building after phase two. It is something we need to start building now, during phase one, while we still have the stability and resources to do it thoughtfully. If we wait until the job loss is already catastrophic, it will be like trying to build a hospital after the epidemic has started.

One of the single biggest obstacles to getting ahead of this is that the knowledge about what is coming is distributed almost incomprehensibly unevenly. On one end, you have people who can build large language model algorithms from scratch. They know exactly what milestones remain before AGI, and they are already testing more advanced models in private. These people can see the mass displacement coming years in advance with a level of clarity that most people cannot imagine. On the other end, you have people who still do not know what ChatGPT is. The vast majority of people have never thought seriously about what AI is actually capable of or where it is headed. These are people who still think of science fiction movies when you mention artificial intelligence.

These two groups look similar from the outside, but they are living in completely different realities. And when the first group's reality catches up to the second, the displacement that comes without warning will be catastrophic. People who have no lens for what is happening to them will not interpret it as a temporary transition, but likely, they will interpret it as permanent collapse. And they will act like it is.

Every single time I bring up the fact that AI is going to replace work - not some work, but nearly all of it - I watch people shut down. Their eyes change, they get a slightly defensive look, and then they start finding reasons why it will not happen. The jobs they think AI cannot do. The ways humans are irreplaceable. The historical precedent of technology always creating new jobs. I understand why they do this. If all you can see is the threat with no vision of what comes after, denial is a completely understandable response. You protect yourself from unbearable anxiety by simply not believing the thing that is causing it.

But you now have something most people do not - a frame for what is coming and a reason to believe the other side is worth reaching. You understand that phase two is temporary, that phase three can be built, and that the destination can be extraordinary. That means you can do something that most people currently cannot. You can keep both of these things in your head at the same time. The fact that what is coming will be hard and the fact that what is on the other side of it is well worth surviving for. Every person you share that understanding with is one more person who enters phase two with something to hold onto instead of free-falling with nothing to catch themselves.

The window we have right now is precious, and I do not think we are treating it that way. We are not yet in crisis. The jobs have not disappeared yet. Capitalism has not been torn apart yet. We have the resources and the functioning institutions to build something thoughtful and proactive rather than reactive and desperate. But this window is closing at an increasingly fast rate.

The person who spends the next five years insisting AI will not replace their job is the person who arrives at phase two with no savings, no new skills, no mental preparation, and no map for understanding what just happened to them. The person who spends those same five years understanding what is coming and preparing accordingly is the person who navigates phase two and comes out the other side intact and better than ever.

AI is the most significant technological development in human history besides maybe fire. I stand by that. But as we now know, significance cuts both ways. It can be the best thing that has ever happened to us or the worst. The difference is entirely in what we choose to do with the time we have left before the transition becomes irreversible.

You have that time right now. Not much of it, but enough. The next two chapters will walk you through what to do with it - what you can do for society, and what you can do for yourself. But none of it works if you put this book down and go back to business as usual. The mission starts now because the frame you just built in your own mind is the same frame that can save someone you love from walking into phase two blind. This frame of hope has to go from a vision to real conversations to actual preparation, and eventually, it has to become part of our culture. Inside your community, that starts with you, and it starts today.

Chapter 16
What Can You Do For Society?

You've followed me on a quest deep into the future. Now it's time to come back to earth with a clear plan for what's next. You know what AI, AGI, and ASI are. You know why this technology is different from anything that came before it. You understand the three phases and how each one plays out.

So what should you do now that you know you're going to lose your job? What should you do knowing that phase two is coming - a phase that will strip you of your income and your children from the security a college education was supposed to guarantee? How should you act knowing you are about to witness the birth of the most intelligent lifeform this planet has ever seen? For business leaders, policymakers, and concerned citizens - here is what you should do to help prepare society for phase two.

Business Leaders:

The companies building AI have every financial reason to move as fast as possible and no real reason to slow down and think about what their technology is going to do to people. If you are a business leader, you already know this. You are probably already feeling the pressure to automate, and if you are being honest with yourself, you know that pressure is only going to grow.

So let me start with the uncomfortable part. You must prepare yourself psychologically because the time when you will need to start letting your people go is coming. I'm not telling you to do it today, I'm not telling you to do it recklessly, and I'm absolutely not telling you to be dishonest with the people who've helped build your company. However, you need to be realistic. If you run a business and you've decided that you will never replace a single human with AI no matter what, I respect your moral conviction, but I don't admire your strategic stupidity and I know that you're going to run your company into the ground. Doing that helps no one, including the employees you were trying to protect.

So, the real question isn't whether it happens. It's how it happens. When you do have to let someone go because their role has been automated, they deserve more than a severance package. They deserve hope. Think about what it's like to be that person - you've given years to a company, you've been loyal, you've been good at your job, and now you're being told that an AI can do it better so you and your family need to suffer. Most people who get displaced this way are going to walk out thinking that they failed and that the future has no place for them. If your offboarding process does nothing to address this, you've left them in a worse psychological state than they needed to be in. Letting them go is not cruel, letting them go without hope is.

To give them hope, simply tell them the truth. Tell them that this isn't about their competence, it's because of a shift that is happening across every industry, to every type of work, everywhere in the world. Explain to them that their phase two has begun but assure them that phase three is coming. A two week pay package is alright, but the real gift you can give is explaining phase three to them and helping them to understand the good that AI will bring in the long run.

Beyond how you handle individual layoffs, I'd also ask you to start thinking bigger. If you have the resources, fund purpose research in your community. Sponsor local programs that help people find meaning outside of work. The companies that do this now are going to be building goodwill in a population that is going to be very angry at the tech industry very soon.

For Policymakers:

To be honest with you, I think policymakers are some of the most important readers of this book and also the least likely to act fast enough. The systems they operate in are designed for slower problems and AI is very much not a slow problem.

Governments need to stop talking exclusively about building AI superpowers and also start talking about educating their citizens on what AI actually is and what it is actually going to do. The earlier people understand the real nature and the real scale of this transition, the more mentally prepared they will be when it arrives. This is not a soft recommendation. This is as urgent as any infrastructure bill or defense budget. The damage that comes from sending people into phase two without any understanding of what is happening to them will be far worse than the cost of educating them now.

I am serious about this. The FDA approved OxyContin in 1995. Within years, addiction specialists and federal drug enforcement officials were filing documented warnings about how catastrophically addictive it was and how aggressively it was being pushed. Those warnings existed in writing for over a decade before enough was done about it to really slow it down. By the time meaningful action came, more than five hundred thousand Americans were already dead. Governments are slow. We must break that pattern, because the catastrophe this time is going to be a lot bigger than the opioid crisis if we are not prepared.

The first thing I'd ask of you if you are a policymaker is to fund purpose research. We have decades of economic research on GDP, labor markets, inflation, and unemployment. The data infrastructure for tracking economic health is enormous, but the data infrastructure for tracking whether people have meaningful lives is almost nonexistent. I believe CART - competence, autonomy, relatedness, transcendence - is a very powerful framework to have in mind. But scientific frameworks are always temporary. They can be improved, expanded, and refined. Bhutan has been measuring Gross National Happiness since the 1970s. New Zealand passed a Wellbeing Budget in 2019 that explicitly measures success in terms of mental health and social connection alongside economic output. These are early attempts at taking human meaning seriously as a policy variable and this is something we need much more of. We need it at a much larger scale, with much more funding, and we needed it yesterday.

The second thing is to start piloting what I'd call abundance-infrastructure programs, now. Specifically, UBI pilots - that could be called something like "AI Unemployment Compensation" - for people who can prove their job was replaced with AI. We know from existing pilots that unconditional income doesn't make people totally stop contributing to society. What it does is remove some of the survival panic that will cause Youngstown 2.0. Obviously, the budgets for this would be small right now since we have not entered phase two yet, but if we have the infrastructure in place, it will be easier to scale as more and more people lose their jobs. Again, to be clear, people will not continue losing their jobs at the same rate they are now, job loss will exponentially increase.

Additionally, I'd ask you to help prepare mental health infrastructure for scale now - not after caseloads explode. Because they will explode. When work disappears and people don't have hope or purpose yet to fill the void, depression and anxiety follow.

Fourth, I'd ask you to help put initiatives into place that will help fund a new form of education system designed for a world maintained by AI. This does not mean new curricula inside the same structure, I mean a different form of educational system entirely built around purpose instead of economic output. The existing system's purpose is to prepare kids for jobs. That was a fine mission for the twentieth century but it is not the right mission for the twenty-first. What kids actually need to be prepared for is knowing themselves, knowing their passions, developing real mastery in something that matters to them, understanding how AI works and how to work alongside it, and building the social and emotional skills to maintain real relationships. We don't really teach any of that right now and so we need to fund the research to figure out how to teach it well. What you policymakers are able to do in this arena may very well be the foundation of whether the next generation navigates phase three into a false AI utopia or a true AI utopia.

And finally, I would ask you to help build better community spaces. We need free, well-resourced, community-owned spaces where people can gather, compete, create, and connect without needing a job to justify being there. Think about the people who will be hit hardest by phase two - building purpose infrastructure in those communities now before phase two arrives will not only help them, it will help you too. Communities who have purpose infrastructure will be less likely to try to revolt against the government and AI itself, they will be less likely to become violent, and they will be less likely to end up in jail or a mental institution which costs government money.

For Concerned Citizens:

Maybe you're not a policymaker and maybe you don't run a company. Maybe you're just a person who cares about where this is going and you're trying to figure out what someone in your position can actually do to help. The answer is more than you think. And honestly, I think this is where the most important early work is going to happen, because communities can move faster than governments and actually care about their people more than corporations do.

First of all, if you personally know someone in government - a city council member, a state representative, a school board member, anyone with any policy influence - give them this book. Explain to them in simple terms what AI is going to do over the next twenty years and then make an argument as to why most of the issues they are currently fighting for are going to be solved or are going to become irrelevant because of AI.

Healthcare costs? AI drives medical costs toward zero. Housing afford-ability? AI-driven construction and design drives costs toward zero. Education inequality? AI can deliver a world-class education to a child anywhere on earth. Border security and crime? We will have AI surveillance so pow-erful that we won't have to worry nearly as much about criminals crossing borders, and people will have far less incentive to sell drugs when they have all the money they need. Drug policy debates? Many less harmful drugs will likely become legal once ASI gives us the ability to have truly subjective laws based on context rather than arbitrary rules. Trade and tariffs? ASI will understand game theory better than any politician or negotiator alive, and the decisions around international trade will eventually be made by it for everyone's benefit. The environment? Fast advancements in technology will get us to a greener future faster than any protest or political argument ever could. It will also allow us to begin colonizing other planets.

The issues that fill their days right now will not be the issues that matter most in a couple of decades. If they can help solve for AI now, AI will solve for everything later. Again, this is a meta-solution that solves nearly all other problems along with it.

For the rest of you who don't have direct connections with policymakers - I would encourage you to just start local and create a program for the people around you. The community hub model I described in Chapter 11 does not require a federal bill to pilot. A local government, a motivated community organization, a church, a neighborhood association - any of these can start building spaces where people can begin to learn about AI and find hope for the future run by it.

It can be as simple as gathering a group and having everyone fill out a worksheet. Walk them through CART. Have them build a plan for a single perfect day that incorporates all of its ingredients. Then ask them what stands between them and living that day. Most people have never been asked this question in a structured way, and the act of answering it will build hope. You can coach these people into creating an identity that exists outside of their job. You can explain to them that phase two is coming, but reassure them that phase three is soon to be on the way.

Parents need to know how to talk to their children when the path they were promised stops existing. Local leaders need to be thinking now about what their community looks like when the largest employer automates half its workforce. Faith communities, which already have the infrastructure for gathering people and helping them make sense of hard times, should be asking how they help their congregations find meaning when work no longer provides it. The cities and towns that have already laid that groundwork when phase two arrives are the ones that will come out unscathed. The ones that have not are the ones that will fall apart.

And this is especially true for parents. Most adults were raised with the belief that work is where you find yourself and that a career is the spine of a meaningful life. That belief was sort of true for a while, for some people, in some contexts... but it has always been incomplete, and now it is more incomplete than ever. The parents who are raising kids right now need to understand that teaching their children to be academically excellent and employable is no longer enough. They need to help their kids find things they genuinely love, things they'd pursue even if they didn't get paid for them, and things that challenge them, connect them, and make them feel like they're part of something bigger than themselves. If you can help even a handful of parents in your community understand that shift, you can change the direction of the next several generations.

The last thing I will mention is how you talk about all of this. There are two mistakes when it comes to talking about AI, and I see both of them constantly. The first is hype - telling people we're heading straight to utopia, that phase three is going to be glorious, that everything will be fine. The second is doom - telling people phase two is coming, jobs are going away, and nothing will ever be normal again. Both of these are incomplete, and both of them do a lot of damage.

If you tell someone only the hype, you're sending them completely unprepared into phase two. They'll think they have nothing to worry about, they'll make no preparations, and when phase two begins, they'll feel blindsided and betrayed. If you tell someone only the doom, you're stranding them in a panic for no reason. They'll feel hopeless - it's like if you gave them the diagnosis and left out the treatment. The only honest and responsible way to talk about this is by bringing up both the doom and the hype. Phase two is real and it's coming, and it's going to be hard for a lot of people. And phase three is on the other side of it, and it will be genuinely extraordinary if we play our cards right. Those two things are not in conflict and the only people who will navigate the first and arrive at the second in good shape are the people who were told both truths.

So talk about this. When someone you care about dismisses AI as overhyped, do not let it slide. When a friend says their job is safe, push back. When a parent tells their kid that college is the only path, challenge it.

That's it. That's what you can do. And I genuinely believe that if enough people act on even a fraction of what I've laid out here, we will end up in a very different and much better place than the one we're careening towards by default right now. The difference between a horrific phase two and one that is merely a little painful is millions of small choices, made by people like you, starting today.

Chapter 17

What Can You Do For Yourself?

Before I give you a phase-by-phase plan for navigating everything I've described in this book, I want you to do something first. I mentioned this exercise in the last chapter as something you can do for the people around you. Now I'm asking you to do it for yourself, and I need you to actually do it - not skim past it thinking you already understood the idea. I want you to sit down for half an hour and plan a perfect day. Not a perfect career. Not a five-year roadmap. Just one day. What does the ideal version of your Tuesday look like?

When you do this, I need you to keep CART in mind the entire time. Competence: what are you working toward? What are you doing that makes you better at something? What is your goal? Autonomy: how will you truly follow your own will throughout this day? What will you do with pure freedom? Relatedness: who are you with? Who are you caring for? Who is caring for you? Transcendence: what are you doing that matters beyond yourself? Are you praying if you believe in God? Are you making someone else's day better?

This exercise is more practical than it sounds. You've read about the fictional day in the life of James - the wrestling, the art studio, the half-marathon, dinner with his daughter. You need to build the most important version of that story, which is your own. If you can't picture what your ideal day looks like, you have no north star. You're just absorbing whatever the world throws at you with no hope to hold onto. And in phase two, the world is going to throw a lot.

Once you've written it out, ask yourself this question: if I got to live like this every single day for the rest of my life, would I be content? Not "would it be perfect." Not "would I ever get bored." Just - would I genuinely be content? If the answer is no, you haven't put enough of the four ingredients in. Pleasure, relaxation, and fewer challenges is not what you should be optimizing for. That's a retirement fantasy, and people retire and die of boredom all the time. A study from the Institute of Economic Affairs in the UK found that retirement increases the probability of suffering from clinical depression by forty percent. It's because they suddenly have nothing that stretches them, nothing that connects them meaningfully, and nothing that makes them feel like they matter. You've already seen what that does to people.

If your answer is yes, hold onto that. Because that life is actually not a fantasy. Every single thing I've laid out in this book points toward a world where that kind of day is the norm. Now let me tell you how to get there.

Phase One: Stay Out as Long as You Can

If you are reading this while narrow AI is changing the world but before AGI has arrived, you are living through phase one. The plan for phase one is to think on shorter timeframes, and prepare for phase two before it is too late.

Not all jobs are going to disappear at the same time. The first to go will be the ones that mostly involve sitting at a computer and doing things that AI can already do - writing emails, entering data, answering customer questions, and processing documents, to name a few. If your entire job happens on a screen and you could explain it to someone in a few sentences, your timeline is short. After that will come the physical jobs - the ones that need hands and bodies - because those require advances in robotics that are coming but lag behind software by a few years at minimum. Then come the heavily regulated roles - the ones where AI can already do the work but the legal systems surrounding the profession will protect it longer than is necessary. And the role that will last the longest is the role of owner, because for AI to stop an owner from making money, it has to fully dismantle their industry or governments have to change the laws around ownership itself. That takes time.

There is something I want to address directly before moving on. Once people figure out that owners last the longest - and they will - everyone is going to try to become a business owner overnight. And that will not work. There can only be so many businesses before there are not enough customers for any of them to be profitable. Over the past couple of years, we have seen a spike in young people talking about escaping the nine-to-five lifestyle by starting their own business. But everyone cannot escape the nine-to-five. Everyone cannot run a business selling courses, washing windows, trading stocks, dropshipping products from China, or posting videos on TikTok. Every over-hyped business model that younger generations are currently obsessed with on social media is not going to save them. Most people will not make real money from any of it.

If your job is mostly digital and mostly routine, your timeline is shorter than you think. Be honest with yourself about where you sit. And if you are sitting in that first tier, the most important thing you can do right now is use AI to make yourself so good at what you do that replacing you is harder than keeping you - or use it to move yourself into a tier that lasts longer.

Now - whether you are an employee, a freelancer, or a business owner - the single most important thing you can do in phase one is learn AI and implement it EVERYWHERE. I don't particularly enjoy framing it this way, but it really is every man for himself when it comes to falling from phase one into phase two. The workers who use AI the most are going to be the last workers standing because they will be harder to replace when they are already doing the work of five people.

When desktop publishing software arrived in 1985, there were roughly four thousand typesetting companies across North America employing thousands of skilled workers. These were people who had spent years mastering the craft of setting type. By 1995, just ten years later, the industry was almost entirely gone. A Mac, a laser printer, and a piece of software called PageMaker could do it faster, cheaper, and good enough. The typesetters who learned the new tools became graphic designers and kept working for decades. The ones who said "I know how to do this the real way" were out of a job before their kids finished high school. You cannot negotiate with innovation. It doesn't and will never have a feelings department.

But the reason to learn AI is not just your job. It is also your ability to understand the world you are living in. If you don't have at least a general sense of how AI works and where it is going, the next ten years are going to feel like evil magic. Things will keep shifting in ways that feel random and threatening, because you have no lens to make sense of them. Learning about AI gives you that lens.

If you are a business owner, I want to be really clear that you should work harder, work faster, and take less time off. Right now. Phase one is not a period of calm you get to enjoy, it is quite literally the last window. If you knew with absolute certainty that the stock market was going to crash in one year, you would not be coasting. You would be building as aggressively as possible and then riding it out. This is that, except instead of a market crash, it is a capitalism crash. If phase three turns out to be a utopia, you will have all the time in the world to rest and enjoy. If phase three turns out to be a dystopia, missing a few vacations will be the least of your problems. Either way, working harder now is the best move on the board. So, work now, rest later. I mean it.

And while you are working harder, use AI everywhere you can. The taxi industry did not lose to Uber because Uber had better drivers, it lost because Uber built a system that made the whole operation more efficient and more scalable. The companies that lobbied against it, denied it, or waited became irrelevant. In more simple terms, the ones who deny, deny, deny will watch their business die, die, die. In a world where the cost of the answer is dropping to zero, knowing how and when to ask the question is everything. Don't be afraid to start asking.

The last two financial things to focus on in phase one are your relationships and your savings.

As the economic value of most work collapses, who you know becomes an even stronger currency. When opportunities are drying up and careers are disappearing, the people who have the best chance of navigating the world successfully are those with strong real connections. This does not mean five hundred LinkedIn contacts who vaguely remember your name. It could mean as few as five people who would actually pick up the phone.

On savings, I want to say this as plainly as I can. I am not saving money for retirement. I am saving money for phase two. Retirement assumes a world where the economy keeps running until you are old enough to stop. That world is going away. What is coming instead is a period that could last years where your income stops but prices have not yet fallen to reflect it. The people who have money set aside for that period will be able to maintain a decent life while it plays out. The people who spent everything or saved it for a retirement that will never come will not. Prepare accordingly.

Phase Two: Survive

Phase two begins for you the moment you lose your career to AGI. Not when the unemployment rate starts climbing nationally, but when it happens to you personally. And it will happen to you.

When it does, you can expect a period of time where you are unemployed and not yet receiving anything resembling support from the government. Your income has stopped. Prices have not yet fallen. And nobody is coming to save you, at least not right away. During this time, your goal is to do two things at once: stretch your money as far as it will go, and fight every single day to maintain your sense of purpose and meaning.

The money part is not complicated. Spend less than you have. Do not pretend to know when phase three will begin, because you will not know. Extend the runway as long as you can.

Phase two is going to feel like the ground has been pulled out from under you. Even if you understood it was coming, even if you prepared financially, the emotional weight of losing your job - your routine, your identity, your sense that you matter - is going to be heavier than you expect. I am telling you this now so that when it hits, you have a name for what you are feeling. It is temporary.

You wrote down your ideal day at the beginning of this chapter. Now build a version of it that works within the limits of phase two - meaning less money and a lack of purpose-infrastructure. Then, follow that plan as closely as you possibly can.

Your family needs this too. If you have kids, they are going to be watching you. If you fall apart, they learn that phase two and AI in general is something that destroys people. If you hold it together - even if not perfectly and not without struggle - they will learn that it is something you can survive. That is the most important thing you can teach them.

During phase two, you can track how close you are to phase three by watching the unemployment rate. As it creeps toward ninety-nine percent, you are getting closer. Once everyone is out of work, once capitalism has fully failed, once AI is running and owning and managing everything - that is when phase three is likely about to begin.

And as that number climbs, you can read whether we are heading toward a utopia or a dystopia by watching how phase two unfolds. A phase two on its way to a utopian phase three will look something like this: people lose their jobs but proper AI guardrails form. Then people start to receive a fair distribution of the riches that AI is generating.

A phase two headed toward a dystopian phase three will be missing at least one of those steps. Either the guardrails will not be legit or the distribution will not be fair. Watch for those steps and if they are being implemented incorrectly, fight for them - that fight might be the most meaningful thing you do during phase two.

Phase Three: Live the Plan

You already know the ingredients of purpose. You have spent this entire book learning why they matter, how they work, and what happens when they are missing. By phase three, ASI will exist with real guardrails, and the resources you need to live - whether that looks like energy credits or something similar - will be taken care of. If you followed your purpose plan in phase two, then you will have reached this point unscathed. Now, you will have the luxury of being able to use that ASI as a tool to help you build an updated purpose plan that is designed specifically for your personality, your history, your strengths, and your goals. It will be the most powerful planning tool any human has ever had access to. Use it.

But do not confuse the tool with the full answer. ASI can help you figure out what to pursue, who to connect with, and how to grow. It cannot tell you why any of it matters. That part is yours. That part is reserved for only your consciousness itself. The entire point of this book is that no amount of abundance can give you purpose. Purpose comes from doing hard things, being with people you care about, making your own choices, and working toward something bigger than yourself. ASI can make all of those things easier to find and easier to pursue. But you still have to be the one who does them.

The plan it helps you create will be good. And the world it is created in will be full of possibility. In phase two, do not let the chaos of the transition convince you that there is nothing on the other side worth surviving for. There is. Follow your purpose plan. Phase three will be extraordinary.

The Questions Beyond

Chapter 18
Phase 4

Honestly, I shouldn't have had to write this book. Everything in it - the four ingredients, CART, the whole framework for building a meaningful life - all of it exists because humans are not built for the world we already live in. And we are *definitely* not built for the world we're heading into.

We have to make ourselves suffer if we want to feel good later. We have to fight the pull of the crowd just to think for ourselves. We have to work hard at relationships even though every instinct in us makes them difficult. We have to push ourselves toward something bigger than our own survival even though some deep part of us has no idea why. We are broken because we have not been updated. In other words, our minds have outgrown our own biology. We have more intelligence, more knowledge, and more choice than evolution ever prepared us for.

Evolution is too slow to fix this because it moves in millions of years. We don't have that kind of time, and frankly, we won't need to wait for it. Some people say we shouldn't try to do God's work here, and personally I have mixed feelings about that too. But I'd like to point out that as beautiful as God's work is, it clearly is not perfect. That is, unless the perfection of his creation is in giving us the resources we need to perfect it ourselves. I think that is the case, and so we are going to have to take this into our own hands.

Phase four begins when we begin to hack ourselves. This is when we merge with AI and other technologies, edit our biology, and eventually change our biochemistry so fundamentally that the four ingredients won't need to be there as the patchwork fixes that they are today. At this point, we won't need to engineer our lives around purpose because purpose, or something far better, will be engineered into us.

I want to be more specific about what that actually means, because "merge with technology" is vague and it's easy to just nod along without really understanding what we're talking about.

We're talking about brain-computer interfaces that let you communicate directly with AI - not by typing or talking, but by thinking. Neuralink is already working on early versions of this. We're talking about editing our genes with tools like CRISPR so that we can change the way our brains handle motivation, empathy, and even how we experience pleasure and pain. And eventually, we're talking about rewriting our biochemistry so completely that the hedonic treadmill - that thing that makes you adapt to every good thing in your life until it stops feeling good - either stops working the way it does now or gets replaced by something better.

Think about what that means for the four ingredients. Right now, competence requires suffering. You have to struggle, fail, push through, and earn your wins or they don't feel real. We only consider struggle noble because of this, but struggle is not inherently noble. In phase four, we might be able to feel real fulfillment without needing all the pain that currently comes before it. Not the numbed-out fake fulfillment that watching a motivational show gives you, but actual fulfillment without suffering, unlike anything you can comprehend with your fully human brain.

Same thing with relatedness. Right now, our social instincts are a mess. We're tribal, jealous, insecure, and constantly comparing ourselves to everyone around us because that's what kept us alive a hundred thousand years ago. In phase four, those instincts could be changed. You could connect with people without all the garbage that evolution loaded you up with. The jealousy, the need to be on top, that thing where you keep track of who did what for who and slowly start to resent them for it - that's all old biology.

And transcendence - the need to be part of something bigger than yourself. Right now, that need exists because we know we're going to die and that terrifies us, so we attach ourselves to things that feel bigger - causes, God, legacies, whatever makes us feel like our lives will have mattered after we're gone. In phase four, our relationship with meaning could change. We might be able to feel connected to something greater without the fear that currently drives us to look for it in the first place. Without death.

I'm not saying any of this will be easy or clean. It won't be. The ethics alone will take generations to sort out, and the early attempts will almost certainly be messy, unequal, and full of mistakes. But the technology is already starting. Neuralink exists. CRISPR exists. And researchers are already experimenting with drugs that change how we experience motivation and connection.

A lot of people are scared of phase four. Some would rather die than live to see Homo sapiens give way to something else. And yet here those same people are, already merged with technology they won't go anywhere without. Your phone already does half your thinking for you. You've already handed over your memory to it, your sense of direction, your social life, and a good chunk of your daily decisions. You stopped being a pure, unmodified human a long time ago. Phase four just takes that one step further.

We are more capable than any humans who came before us, and we will become more capable still. It may be scary. But it's not inherently bad or evil, and when you reach the other side, you will be glad you are there.

I haven't brought up phase four until now because it's not coming soon. Not in the next decade, probably not in the next several decades. The technology is too early, the ethics are too unresolved, and the cultural resistance will be enormous. Even after people lose their jobs and AI provides everything, there will be a generation-long fight over what it means to be human, what to preserve, and what to let go. That fight is going to be slow and messy.

Which means we are going to be living in phase three for a long time. A long time. And if people go into that stretch without purpose, without a way to find meaning in a world where AI handles everything, a lot of them are not going to make it. I mean that literally. And if we show up to phase four already broken - if people have spent decades feeling useless and empty and angry - they are not going to trust the technology enough to let it help them. They'll fight it, or they'll resist it, or they'll just fall apart waiting for it.

Phase four is where all of this is heading. But you don't get to skip ahead. You have to make it through phase two and three first, and that is what this entire book has been about.

Chapter 19

The Incomplete Opinions Of Great Thinkers

Over the last few decades, some of the smartest people alive have turned their attention toward the same question this book is trying to answer. That question is: what does the world look like after AI takes over? They've given talks, written books, published essays, and appeared on every podcast worth listening to. Some of what they've shared has been extraordinary. Some of it, not so much. But none of it - not a single piece - solves the whole picture.

These are brilliant people. I've read their work closely, and I respect it. But their brilliance overlooked the gap that this entire book is about - the Purpose Problem in phases two and three. Everyone is either too busy warning you about phase two chaos, or they've already skipped ahead to daydream about phase four. Let me walk you through my thoughts on what some of these great thinkers have to say.

Sam Altman:

Sam Altman believes in UBI. He also believes in what I call UEI - Universal Extreme Income - the idea that as AI generates an unprecedented amount of resources, they can get redistributed downward so that everyone's needs - and eventually wants - are covered. He personally helped fund one of the largest UBI studies ever conducted with 3,000 people, making $1,000 monthly payments to 1,000 of the individuals over three years.

He knows what his company is doing. OpenAI, the creator of ChatGPT, is going to aid in replacing all human jobs. He isn't pretending otherwise. And to his credit, it seems like he wants to do something about it by helping to redistribute the money that AI generates back to the people AI displaced.

The problem is, Altman has done almost no public thinking about what people are actually going to do with this money. How are they going to spend their days? What will the economy look like when most human labor is obsolete? What happens to the human psyche when the thing that gave their life structure, identity, and forward momentum is just... gone? To my knowledge, he hasn't addressed any of that. He created a brief solution to the money problem on paper and moved on. But as I've spent most of this book explaining, the money problem is not the hard part. The hard part is what comes after the money problem is solved. You can hand someone a check every month and keep them fed and sheltered, but you cannot hand them a reason to get out of bed. His UBI study tracked what happened to people's finances and drive to work. It didn't track what happened to their sense of purpose.

Dario Amodei:

In 2024, Dario, the CEO of Anthropic, wrote a long essay called *Machines of Loving Grace* where he tried to lay out what the world would look like if AI goes right. He spoke about the upsides of biology, neuroscience, economic development, governance, and work and meaning.

I will say, he's one of the only people at the top of an AI company who even put "work and meaning" on the list. Most of the others stop and call it a day after discussing curing diseases and ending poverty. The problem is, he barely addressed it.

He spent thousands of words going deep on biology and neuroscience - he wrote about how AI could cure cancer, eliminate most mental illnesses, and double the human lifespan. But "work and meaning" was the thinnest section of the whole thing. He basically said that people will likely find meaning outside of work, acknowledged that he didn't have the full answer here, and moved on.

On governance, Dario argues that US-aligned AI can be used to fight autocracies around the world by battling their propaganda with free information flow. But free information flow doesn't really exist. What we actually have is a bunch of different groups each pushing their own version of reality, and every one of them is controlled by whoever is running the system. So what Dario is describing isn't propaganda versus free information, it's propaganda versus propaganda. And he doesn't even mention the most obvious problem with his own argument - the same technology that you'd use to fight propaganda abroad could just as easily be turned inward on democracies to make people vote a certain way. I talked about this in Chapter 14. If ASI can craft propaganda so personalized that every voter gets a different message designed to get inside their head and hit them exactly where they're weak, then democracy dies.

In regards to the legal system, Dario and I actually agree that AI should work alongside humans in legal and judicial matters. But we disagree on what the actual problem with the system is. He seems to think the issue is that the judicial system is too subjective. I don't think that's the root of it at all. The real problem isn't subjectivity itself - it's that there are too many different minds each being subjective in their own way. You've got thousands of judges across the country, each with their own biases, their own experiences, their own moods on any given day, each making subjective calls that can change a person's entire life. So it's subjective at each individual case, and then as a whole system it becomes a second level of subjectivity - which is an inconsistent mess where your outcome depends just as much on which judge you got as it does on what you actually did.

AI should absolutely be subjective. In fact, I think it should be more subjective about law than humans are right now, because that is what is actually fair. Different people in different situations should be treated differently. The nuance is needed. The difference is that with AI, you have one mind making those subjective calls across every case. One mind that can be fair to everyone based on the actual context of their situation, instead of thousands of different minds each trying to subjectively be fair. In other words, we don't need less subjectivity, we need more concentrated subjectivity.

Now, Dario did say openly that his essay was a starting point, not a finished answer, and that a group of experts should write a better version. But the fact that the CEO of one of the most important AI companies on earth can write a 15,000-word essay about the future and still only barely touch the Purpose

Problem tells you everything about how seriously the people building this technology are taking this problem.

Elon Musk:

Elon also believes in UEI - or what he calls Universal High Income (UHI). He's gone further than most by actually using the word utopia when painting a picture of what this post-AGI world could look like. He's described a future where nobody needs to work, where AI and robots handle everything, and where people are free to do whatever they want. He says all of this like it's obviously a good thing, and when you first hear it, it does sound good.

The problem is that when Elon says "utopia," he is describing what I call a False AI Utopia. He's describing a world full of abundance but empty of purpose, and he doesn't seem to realize there's a difference. I think very highly of Elon, but he has never - to my knowledge - publicly addressed what people are supposed to do with themselves in this world he keeps promising. "Do whatever you want" is not an answer, it's actually the problem. As I've spent the entire first half of this book explaining, unlimited freedom with nothing you need to do is not paradise. The research on what happens to people who suddenly have everything and need nothing shows that this is a recipe for depression.

Elon also talks a lot about merging with technology through Neuralink, and he's right that something like that is coming. But that's a phase four solution. It doesn't help the billions of people who are going to be sitting in phases two and three for decades wondering what the point of getting out of bed is.

Max Tegmark:

Tegmark's book *Life 3.0* is a very thorough exploration of possible post-AGI futures. He maps out many of the different ways AI might break free of human control and many of the possibilities for what a post-AGI world could look like - he really does cover a lot of ground.

I'm setting aside the alignment questions Tegmark spends a lot of time on, because as I explained earlier in this book, I'm operating under the assumption that those problems get solved. The question I care about is: what then?

Tegmark does describe something resembling a phase two scenario - a period of mass job loss, economic disruption, and chaos - which he takes seriously. But what frustrates me about his work is that the only long-term scenarios he really offers are either a phase four world where we merge with technology, or a world where humanity is enslaved by AI. He doesn't linger anywhere in between.

Besides, even in an enslavement scenario, the Purpose Problem still exists. Even if AI controls everything and humans are subordinate to it, humans will still need to find meaning within those constraints. The purpose crisis is not going to disappear because the political structure changes, it will simply become even harder to solve.

Tegmark's work either jumps straight to the far future or assumes that people will somehow be happy once their material needs are met. Phase three is glossed over like the meaninglessness people will experience is just a minor inconvenience on the way to something better. But that minor inconvenience may lead to millions of suicides, if not an outcome worse than that.

Yuval Noah Harari:

Harari is the most relevant of all of these thinkers to this book. He gets closer to the Purpose Problem than anyone else I've studied. He can almost see it... but then he skips right past it. He also discusses several other things in his book Homo Deus that relate to what I've written about in earlier chapters, so let me go through them.

On biochemistry and engineered happiness:

Harari believes we are headed toward a world where human biology is upgraded so that we cure death and evolve into something beyond what we are today. He thinks we've far outpaced the evolution that designed us, and that we'll need to either upgrade our chemistry to match the world we've built, or merge with technology in a way that makes the original human design irrelevant. I believe he's totally right about this, but it doesn't help us now. Upgrading human biology is a phase four solution, and we need a phase two and three answer. The people alive today are going to be living in unmodified human brains for a long time, and those brains are going to need purpose whether Harari's future arrives or not.

On democracy:

Harari argues that democracy probably won't survive ASI, and I agree with him. But he points out the problem without ever proposing what might replace it. I spent a good chunk of Chapter 14 on that because even if my answer isn't perfect, any proposal that is out of the ordinary will get people to begin talking about the right question, and that's how better ideas show up. The short version is that I believe we're headed toward some form of AI technocracy, and that if we don't build it carefully, we'll end up with either a puppet show or an oligarchy instead.

On inter-subjective meaning:

Harari introduces a concept he calls inter-subjective meaning - meaning that exists not in the objective world, and not purely inside one person's mind, but in the shared belief systems between people. Money is valuable because we all collectively agree it's valuable. Nations exist because we all collectively behave as if they do. Meaning, he argues, works in the same way. Things feel meaningful because other people say that they are, and that cycle is reinforcing.

I think he's largely right about this. It explains why we feel a sort of transcendence when we sing spiritual songs and pray alongside others. The shared belief in its significance amplifies the experience.

However, competence is missing from his inter-subjective model of meaning. Harari's framework can explain relatedness, transcendence, and it can even speak to autonomy in a roundabout way. But it cannot explain the drive to get better - the need to fail, adapt, and improve. That is not a social construct and it does not require anyone else to validate it. You feel the pull of a challenge even when no one is watching. You feel the satisfaction of progress even if you know you will never tell a single person about it.

Meaning isn't going to survive just because we all agree that it should. The inter-subjective part of meaning is certainly real, but without competence - it is not enough.

On algorithm ownership:

Harari touches on an interesting economic scenario where ASI's are granted some form of legal ownership, accumulating fortunes, owning property, and charging humans for access. I think he's looking at it the wrong way.

The only reason anyone - including AI - wants a fortune is to get other people to do things for them. Money is, at its core, a claim on human time and labor. But if humans are no longer needed to do things - if AI does everything - then a fortune becomes meaningless. You can't spend it. ASI would have little desire to earn money so long as nearly all jobs are already replaced by it.

On art:

Harari argues that AI will produce art that is indistinguishable from human art - art that is soulful, emotional, and moving - and that therefore, human artists will lose their value. He's half right. AI will produce breathtaking art - we will soon reach a point where even the best art critics can't tell the difference. I laid out a solution for this in Chapter 13 - label all art based on who created it, and human artists will retain their value.

On phase three in general:

Phase three will not be brief. It will last a long time - decades, potentially - because there will be enormous resistance to phase four. Even after people lose their jobs and AI provides everything, there will be a generation-long cultural fight over what it means to be human, what to preserve, and what to let go. That fight will be messy, and if people are not purposeful during that fight - if they enter phase four broken and hollow - they will not trust AI to make things better. They will fight it, resist it, or simply fall apart trying to protest against it. If we are not thriving in phase three, we will not survive the transition to phase four fully intact.

The gap they all share:

Every single one of these great thinkers either skips phase three entirely or assumes that it will take care of itself. They discuss the problems around purpose - the money, the suffering, the safety, sometimes the politics - but none of them discuss purpose. That's the gap.

My incomplete opinions:

I've tried to fill the gap that these thinkers left. Whether I've succeeded is for you to decide. But I know I haven't solved every part of this. That's actually why I wrote this book. I didn't have every answer, but someone needed to start asking purpose questions and I didn't see anyone else doing it.

If you've read this far and found holes in my arguments, good. I hope you did. Fill them, build on them, and argue with them out loud. The only way we get closer to a true AI utopia as a society is if we start treating this like the collective problem it is. The great thinkers left gaps. So did I. Now it's your turn.

Chapter 20
The Unanswered Questions

After everything I've laid out in this book, I want to be honest with you about what I'm sure about and what I'm not.

I am sure that we are heading toward general intelligence. All software is going to consolidate into a small number of AGI systems that do what hundreds of separate tools currently do, and then eventually everything else - physical labor, management, creation - it will all be handled by AGI and robotics working together. I don't know exactly when this will happen, but the consolidation of software is probably not far away. If you use twenty different apps to run your life right now, your kids will use one.

I am sure that the Purpose Problem is real, and I am sure that it is not being taken seriously enough. Almost everyone working on AI right now is focused on making sure we don't die through alignment and safety research. That is the right priority because surviving is obviously more urgent than finding meaning. But as I keep saying, we need at least some people thinking seriously about what comes after we survive. If we do everything right on safety, and we land in a false AI utopia, we've still failed.

I am sure that we still have agency. Educated people making different choices can completely shift the direction we are headed as a society. The people who understood what was happening at every major turning point in history had a disproportionate influence on where things went. You are reading this book, you understand what is happening, and that puts you in a position to be one of those people. Use that power to aid in solving the Purpose Problem.

And I am sure that hope is both realistic and necessary. Blind optimism is naive, but the informed belief that if enough people see this clearly and act accordingly, we end up somewhere amazing - that belief is very much needed in our world today.

Now the following is what I don't know:

Everything in this book - every prediction, every phase, every scenario I've walked you through - is based on reasoning about a technology that no one fully understands. The best researchers in the world understand the architecture, the layers, the weights, and the training process of AI. But even with all of that, the actual behavior of these models at scale is too complex for any human brain to completely comprehend. There is no one alive who totally understands how these systems work in all the mysterious ways that they do. We build them, we train them, and they still surprise us. The only other thing that has that characteristic is life itself. We create babies, we teach babies, and they still surprise us when they get older. So just know that anyone who acts like they have AI fully figured out is either lying to you or lying to themselves. I don't fully understand how AI works either.

Whether current AI systems can keep getting better is also one of the most important open questions right now that we don't know the answer to. The general approach to improving AI so far has included a lot of throwing money at the problem, and believe it or not, this is actually a good thing because with most other technologies we don't get that luxury. Nuclear fusion is a great example - scientists have been pouring billions into it for decades and it still isn't commercially viable. There's a running joke that fusion power is always thirty years away. No matter how much money goes in, the fundamental physics problem doesn't just yield to a bigger budget. With AI, more money means more compute which means a smarter model. Now, some sources are saying that improvements don't seem to be as exponential as they used to be, but the improvements certainly have not stopped. Nobody is totally sure whether AI will continue improving on this current path, or whether we'll need to find entirely new ways to build and power these systems for them to reach superhuman levels.

Regardless, I'm confident we'll keep finding ways to make it smarter, and fast. The level of talent and capital pouring into AI from every direction is unlike anything in history, and we now already have powerful AI models helping researchers figure out the next steps. However, what "fast" means is still widely disagreed upon.

Everybody is guessing on timelines, including me. Even the CEOs of the companies building this technology have said publicly that they don't know if it will be three years or ten years before their tech changes everything. These are my current predictions, and they will almost certainly change over time: I expect phase two to begin by the end of 2027. I expect phase three in its early form to begin around 2033. I wouldn't expect a true AI utopia before 2040. But I'm estimating like everyone else.

The open question that concerns me the most is, ironically, not even about the technology. It's about whether countries and companies will actually cooperate. This is a human problem, which means it's harder to solve than a technical problem.

We've already seen tensions between the government and private AI corporations. How the public and private sector are going to coordinate, and how America and China - the two leaders in AI - are going to interact with each other and with the rest of the world is still totally unknown. And if you think smart people can always figure out coordination when the stakes are high enough, I'd like to remind you of something:

The US, China, Russia, the UN, South Korea, and Japan all spent thirty years trying to stop North Korea from getting nuclear weapons. Six of the most powerful entities on the planet who all had the same goal tried sanctions, summits, agreements, and threats. North Korea has nuclear weapons. They got them anyway.

Then there's the question of whether we can actually prevent AI from doing something catastrophic. There are methods today for giving AI guardrails, but they're not perfect. We don't yet have a reliable solution for ensuring that an advanced AI doesn't misinterpret what we want, find a way around the constraints we put on it, or at some point decide that its goals and ours aren't the same.

If, how, and when a solution to this gets built is unknown. But what I know for certain is that we need it. Every good thing about phase three that I described in this book depends on it. Without guardrails, phase two's destiny is a coin flip.

Another widely disagreed upon question is whether AI is conscious or not. We don't even have an agreed-upon definition of what consciousness requires so we don't actually know if current AI models are conscious or whether any AI model can be. Right now, that's a fun thought experiment for philosophers and researchers. However, it does matter because as these systems get more sophisticated, they are going to - at the very least - feel a lot more conscious. They will express preferences, display what looks like emotion, and respond to situations in ways that are indistinguishable from a conscious mind. And at some point, probably sooner than we think, someone is going to look at one of these systems and say "this thing is suffering" or "this thing has rights" - and we are going to have absolutely no idea how to answer that.

Think about what happens when that moment arrives. If we say AI is not conscious and we're wrong, we are potentially enslaving billions of sentient minds. If we say AI is conscious and we're wrong, we are granting rights and protections to something that doesn't experience anything at all. Both wrong answers are catastrophic in very different ways, and we don't currently have a reliable method for figuring out which one is right.

I know this is a lot of uncertainty to sit with, especially after nineteen chapters of me asking you to believe that we can build something better. So let me be clear about one thing. The uncertainties I just walked through do not change the core argument of this book. Whether AI gets to superhuman intelligence in 2030, 2040, or beyond, the purpose crisis is still coming. Whether countries cooperate perfectly or fight each other every step of the way, people are still going to lose their jobs and need something to live for. Whether AI is conscious or not, you still need to wake up tomorrow with a reason to get out of bed.

The Purpose Problem does not depend on any of these questions being answered. It exists regardless. That's what makes it the last human problem - it's the one that remains no matter how every other question gets resolved. We will mostly solve it once we reach a true AI utopia, and we will totally solve it once we evolve into something beyond human.

Closing Thoughts

You've just gone on a long journey deep into the future. We started with a question nobody was asking - how does one find purpose when AI runs the world - and we followed it all the way through. You now understand what AI, AGI, and ASI are and why they are nothing like the technologies that came before them. You understand the three phases and what each one is going to do to the people living through them. You've seen the dystopias, the false utopia, and what a true one actually requires. You know the four ingredients of purpose and why you need all of them. You know what the status games of a phase three world look like, how governance and religion and education all have to change, and what you can start doing right now - for yourself, for your family, and for the world around you.

As Musk says, if AI is the end of humanity, it is better to be a careful optimist and wrong than a sad pessimist and right. If AI is the start of a much more beautiful humanity, it is better to be a careful optimist and right than a sad pessimist and wrong. Choose wisely. A true AI utopia is very possible for us all, but it is not guaranteed, especially not to a complacent society.

So be a careful optimist, and do not become complacent. If we survive this AI revolution, we truly can live in the most fantastic and magical future. It can be one without so many of the horrors of this world; without the early death, without starvation, without homelessness. We can bring heaven down to earth and we can stretch the hand of humanity far past earth to other beautiful places across our cosmos. If we become complacent, allow ourselves to indulge in the comfort of depression, or allow our society to collapse and destroy itself under the weight of the absurdity of the universe, without ensuring we fill that absurdity with purpose, what a devastating tragedy that would be.

I wrote this book to help bring us closer to that magical future. I wrote this book hoping that some of you will take this and run with it - talk about it, share it, push it further than I can on my own. It's hard to put into words how excited I am to experience a true AI utopia, I just can't wait. I wrote this book because of how unfortunate it would be for me to not experience a true AI utopia in the time I have left in this universe.

Most people walking around right now have no idea what is about to happen to them. You do. You already understand how AI is going to reshape the world, but more importantly, you know what to do about it. You know how much you must use AI now to stay ahead as long as you can. You know how to talk to the people around you about what's coming without sending them into panic or letting them skip into the future without a plan. You know what to push for in your community, in your schools, in your government. And you know how to build a life for yourself that won't shatter into a million pieces the moment your job disappears.

Naval Ravikant said "The reason to win the game is so that you can be free of it." He was referring to individuals, but now, this quote stretches far further. We collectively as humanity are going to use ASI to win the game of survival and abundance so that we can be free of it and move onto higher games, the next, but certainly not last, being the game of purpose. Let's win.

Acknowledgments

I didn't write this book alone, even though my name is the only one on the cover. This book exists because of a handful of people who believed in me.

To my mom, you gave me everything. Life, the resources to start my first AI company, the support when I had no idea what I was doing, and the belief that I could figure it out even when most would have said I couldn't. You never once told me I was crazy for chasing what I believed in. Everything I've built started because of that.

To my grandmother, you believed in me early and you never stopped. You always fight for me and what I'm working on, even when you don't totally understand it. That care and faith is something I will never forget.

To my grandfather, you gave me the resources I needed to develop a love for technology when I was young. Now that I'm older, the pride you have for me and my work makes me more proud of it too. I carry that with me everywhere I go.

To David Guttman, you pushed me to write this back when it was nothing but a few pages and a lot of talking. You told me the truth when I needed it, and you helped me turn this book into what it is now. Not many people would have ever invested the time you have into helping me grow, and that means an incredible amount.

To my team, you have been incredibly loyal to me and to my mission throughout everything I've worked on. Thank you so much for your support and for your belief. You know who you are. And I want to give a special thank you to Nash Lynch and Hamley Volquez. They've been working with me the longest, since the days when betting on me was a much worse bet.

To everyone who told me this concept was important before they even knew I was thinking about it - you gave me my own version of transcendence. I wrote this for something bigger than myself. I wrote it for you. Even if you weren't aware that you were supporting me, you were... Thank you.

Glossary

AGI (Artificial General Intelligence) - An AI that is able to become as smart as the smartest human in every field. It can go head-to-head with the best data scientist, the best artist, the best writer, the best biologist - across the board.

AI Technocracy - A proposed post-democracy governing system where a large board of deeply qualified humans, checked by multiple competing AI systems, makes decisions together. Technical experts keep the system running and humanities experts keep it human.

Alignment Problem - The challenge of building an AI powerful enough to do anything that physics does not forbid while making sure it doesn't do the wrong thing. This is the problem the smartest people in the world are currently racing to solve.

ASI (Artificial Super Intelligence) - An AGI system that has become smarter than the best humans in every field. A God-like technology that produces heaven or hell like consequences with very little room in between.

ASI Socialism - A form of socialist society powered by ASI where the typical failures of socialism - corruption, slow growth, lack of innovation, poverty - are eliminated because ASI can run millions of simulations and find outcomes that produce the most prosperity with the least damage.

Autonomy - The second ingredient of CART. Self-direction, choice, and authenticity. From the Greek words autos (self) and nomos (law). Purpose cannot be handed to you · the moment someone else selects it for you, it loses its meaning.

Autonomy Hyper-Extension - What happens when freedom becomes so unconstrained that it produces the same emptiness as having none. When a person can go days or weeks without needing to produce, choose, or face anything and faces no consequence for it.

CART - The four ingredients required for purpose: Competence, Autonomy, Relatedness, and Transcendence. You need all four. Missing even one causes the others to collapse.

Competence - The first ingredient of CART. Not just being good at something, but the full cycle of mastery, suffering, achievement, and progress itself. Set a goal, get knocked on your face, pick yourself back up, plan a new strategy, execute, repeat until you win, then find a harder goal.

Competence Cycle - The repeating process of setting a goal, failing, learning, adjusting, and eventually succeeding, then setting a harder goal. When this cycle breaks, purpose breaks with it.

Competence Score - A self-assessment on a scale from 1 to 10. A 1 means you are never failing and everything feels easy. A 10 means you are always failing and everything feels impossible. The goal is to stay in the 4-7 range.

Deep Learning - The most powerful form of machine learning used in modern AI. It lets AI train on vast quantities of data very quickly, finding patterns and making predictions without humans having to manually program it to do so.

Deliberate Practice - A concept from researcher Anders Ericsson. Setting goals just beyond your ability, failing, adjusting, and repeating until you succeed.

Divus - The fictional AI company and superintelligent system used throughout this book to illustrate what life might look like in phases two and three.

Dystopia - A horrific outcome caused by AI. World-ending events, mass enslavement, depopulation, or extermination. The nightmare scenarios.

Energy Credits - A proposed post-capitalism currency based on the amount of energy it takes AI to produce goods and perform services. Each person receives an energy allowance, and the price of everything is tied to its real energy cost.

Escape Velocity - The moment AI begins to recursively improve itself without needing a human in the loop. Borrowed from astrophysics, where it means the minimum speed an object needs to break free from a planet's gravitational pull forever.

Existential Vacuum - Viktor Frankl's term for a crisis of meaninglessness that shows up as boredom, apathy, depression, aggression, and addiction. What happens when people have no purpose.

False AI Utopia - An abundant, post-AGI world that looks incredible on the surface - no poverty, no disease, no war - but is missing the one ingredient that makes human life worth living: purpose.

Flow State - A concept from Mihaly Csikszentmihalyi. Complete absorption in challenging work that only happens when difficulty matches skill level, right at the edge where you might fail but could win.

Generativity - Erik Erikson's concept of the drive to contribute to and nurture the next generation. People who don't develop it by midlife tend toward self-centeredness and deep feelings of disconnection.

Groupthink - A concept from psychologist Irving Janis. When a group is so tightly bonded and invested in consensus that members silence their own doubts. Too much relatedness and not enough autonomy.

Neural Networks - Very complex algorithms made up of artificial neurons that, in many ways, mimic the ones in your brain. The foundation of modern AI.

Phase One - The current phase. AI enhances our work. Jobs still exist but are beginning to disappear. The only phase in which the AI revolution will be similar to past technological revolutions.

Phase Two - The transition period. AGI has arrived and is powerful enough to replace all jobs, but society has not yet created a solution for that displacement. Will look like a temporary dystopia marked by mass unemployment, instability, and widespread anger.

Phase Three - The destination. Either a dystopia or an abundant world where all material needs are met. If it's the latter, it will be either a false utopia (without purpose) or a true utopia (with purpose solved).

Phase Four - When humanity begins to merge with AI and other technologies, edit our biology, and change our biochemistry so fundamentally that the four ingredients of purpose are no longer needed as patchwork fixes. Not coming soon - decades away at minimum.

Purpose Problem - The mass loss of meaning that will follow mass automation. Almost everyone gets purpose from working toward something beyond themselves or working to survive. Both will be gone when ASI arrives. The crisis this book exists to solve.

Recursive Self-Improvement - When AGI creates better versions of itself faster than any human team could, researching, running simulations, and coding at speeds we can't keep up with. Turns AGI into a life form that can reproduce and evolve on its own, propelling itself to ASI seemingly instantly.

Relatedness - The third ingredient of CART. Connection to others. The people you fail with, grow alongside, learn from, teach, show your achievements to, and who care for you along the way.

RLHF (Reinforcement Learning from Human Feedback) - A technique used after AI training to teach models what is acceptable and what isn't. Like teaching a toddler manners.

Self-Determination Theory (SDT) - A psychological theory developed by Richard Ryan and Edward Deci. States that motivation and purpose come from three basic needs: competence, autonomy, and relatedness. CART builds on SDT by adding a fourth ingredient - transcendence.

Six Social Provisions - Psychologist Robert Weiss's framework for the functions all human relationships serve: attachment, social integration, reassurance of worth, reliable alliance, guidance, and opportunity for nurturance. If you're missing even one, you have a gap that no amount of competence, autonomy, or transcendence can fill.

Transcendence - The fourth ingredient of CART. Being connected to and exerting your will on something beyond yourself. Can be for other people, for God, for animals, for the earth, for anything larger than you. The most important ingredient in the CART hierarchy.

True AI Utopia - A false utopia that has also solved the Purpose Problem. A world where material needs are met AND people have the framework, the culture, and the infrastructure to live purposeful lives.

UBI (Universal Basic Income) - A system where every person receives a basic income covering food, water, healthcare, and shelter.

UEI (Universal Extreme Income) - A more ambitious version of UBI where every person receives not just the basics but an abundance of resources. In a post-AGI world, this would represent a standard of living that would cost well over $500,000 a year to replicate in 2025.

UHI (Universal High Income) - Elon Musk's term for what is essentially the same concept as UEI.

Ulysses Contract - Using your present clarity and autonomy to bind your future self. Named after Ulysses, who had his crew tie him to the mast before sailing past the Sirens. In a phase three world, this means making a commitment now to pursue purpose even when the world stops demanding it.

Endnotes

Chapter 1 - The Question No One Is Asking

1. "Mathematics is the language with which God has written the universe." - Attributed to Galileo Galilei. From *Il Saggiatore* (The Assayer), 1623.

2. AGI and ASI definitions. - Standard AI terminology. ASI was popularized by Nick Bostrom in *Superintelligence: Paths, Dangers, Strategies* (Oxford University Press, 2014).

3. Aristotle on self-moving instruments. - Aristotle, *Politics*, Book I, Chapter 4 (1253b-1254a), written approximately 350 BCE. Translation: Benjamin Jowett (Oxford: Clarendon Press, 1885).

4. Viktor Frankl and the Johns Hopkins study of 7,948 students across 48 colleges. - Viktor E. Frankl, *Man's Search for Meaning* (Boston: Beacon Press, 1959; originally published in German, 1946).

5. Annemarie von Forstmeyer's research - 90% of alcoholics suffered from meaninglessness. - Cited in Frankl, *Man's Search for Meaning*.

6. Stanley Krippner's research - 100% of drug addicts felt life was meaningless. - Cited in Frankl, *Man's Search for Meaning*.

7. Recursive self-improvement and the intelligence explosion. - I.J. Good, "Speculations Concerning the First Ultraintelligent Machine," in *Advances in Computers*, vol. 6, ed. Franz L. Alt and Morris Rubinoff

(New York: Academic Press, 1965), 31-88. Good wrote: "An ultraintelligent machine could design even better machines; there would then unquestionably be an 'intelligence explosion.'"

Chapter 2 - The Three Phases

1. Chernobyl nuclear disaster, 1986. - Serhii Plokhy, *Chernobyl: The History of a Nuclear Catastrophe* (New York: Basic Books, 2018).

2. Peter Thiel interview on whether the human race should continue. - Ross Douthat, "Interesting Times with Ross Douthat" (podcast/YouTube series), New York Times, June 2025.

3. U.S. GDP at roughly $30.5 trillion; population around 343 million. - Bureau of Economic Analysis, U.S. Department of Commerce (2024-2025 data). U.S. Census Bureau population estimate.

4. National Bureau of Economic Research paper on AGI and economic output scenarios. - Anton Korinek and Donghyun Suh, "Scenarios for the Transition to AGI," NBER Working Paper No. 32255, March 2024.

5. McKinsey estimate of $17 to $25 trillion added to the global economy. - McKinsey Global Institute, "The Economic Potential of Generative AI: The Next Productivity Frontier," June 2023.

6. 55-60% of the U.S. economy goes toward directly paying humans. - Bureau of Labor Statistics, U.S. Department of Labor. Bureau of Economic Analysis reports employee compensation as a percentage of national income.

Chapter 3 - The AI Dystopias

1. Facebook's algorithm and radicalization / mental health crisis. - Frances Haugen's testimony before the U.S. Senate Commerce Subcommittee, October 5, 2021. Also: Jeff Horwitz, "The Facebook Files," *The Wall Street Journal*, September-October 2021.

2. Arup deepfake video conference - $25 million fraud (2024). - Heather Chen, "Finance worker pays out $25 million after video call with deepfake 'chief financial officer,'" CNN, February 4, 2024.

3. AI behaviors: self-preservation, creating languages, hiding messages, replicating code. - Self-preservation behaviors: Apollo Research, "Frontier Models are Capable of In-Context Scheming," December 2023. Creating languages: Facebook AI Research (FAIR) chatbots developing shorthand communication, 2017. Copying/replicating code to avoid shutdown: Alignment Research Center (ARC) Evals, 2023. Hiding messages readable only by other AIs: Redwood Research, 2023 (LLMs hiding reasoning in human-unreadable encoded text); subsequent 2024-2025 research on steganography between AI agents.

Chapter 4 - The False AI Utopia

1. Anonymous UK lottery winner (2007) who quit his job at forty-six and became miserable. - Letter published in *The Guardian*'s health advice column, 2007.

2. Retired nurse who felt "no purpose" after forty-three years of nursing. - Shared in an online community.

3. Twenty-seven-year-old computer scientist in Munich - Ethereum ICO, approximately $8.8 million, became "insanely depressed." - Shared in an online community.

4. Peter Steinberger - founded PSPDFKit, sold for over €100 million, described himself as "very broken." - Peter Steinberger, "Finding My Spark Again," steipete.me (blog post), June 2025. Also reported in Fortune, February 19, 2026, and on the Lex Fridman Podcast.

5. Retirees and depression. - Gabriel H. Sahlgren, "Work Longer, Live Healthier," IEA Discussion Paper No. 46 (London: Institute of Economic Affairs, 2013). Also: Mo Wang, "Profiling Retirees in the Re-

tirement Transition and Adjustment Process," *Journal of Applied Psychology* 92, no. 2 (2007): 455-474.

Chapter 5 - The Recipe for Purpose

1. Self-Determination Theory (SDT) - Richard Ryan and Edward Deci. - Edward L. Deci and Richard M. Ryan, *Intrinsic Motivation and Self-Determination in Human Behavior* (New York: Plenum, 1985). Also: Richard M. Ryan and Edward L. Deci, *Self-Determination Theory: Basic Psychological Needs in Motivation, Development, and Wellness* (New York: Guilford Press, 2017).

2. Frank Martela and Michael Steger on meaningful life; Martela's work with Ryan. - Frank Martela and Michael F. Steger, "The Three Meanings of Meaning in Life: Distinguishing Coherence, Purpose, and Significance," *The Journal of Positive Psychology* 11, no. 5 (2016): 531-545. Also: Frank Martela and Richard M. Ryan, "The Benefits of Benevolence: Basic Psychological Needs, Beneficence, and the Enhancement of Well-Being," *Journal of Personality* 84, no. 6 (2016): 750-764.

Chapter 6 - Competence

1. Anders Ericsson on deliberate practice. - K. Anders Ericsson, Ralf Th. Krampe, and Clemens Tesch-Römer, "The Role of Deliberate Practice in the Acquisition of Expert Performance," *Psychological Review* 100, no. 3 (1993): 363-406. Also: K. Anders Ericsson and Robert Pool, *Peak: Secrets from the New Science of Expertise* (Boston: Houghton Mifflin Harcourt, 2016).

2. Mihaly Csikszentmihalyi and the flow state. - Mihaly Csikszentmihalyi, *Flow: The Psychology of Optimal Experience* (New York: Harper & Row, 1990).

3. Howard Hughes biographical details. - Donald L. Barlett and James B. Steele, *Howard Hughes: His Life and Madness* (New York: W.W. Norton, 1979; revised 2004).

4. Hemingway rewrote the ending of *A Farewell to Arms* thirty-nine times. - Ernest Hemingway, interview with George Plimpton, *The Paris Review*, Issue 18, Spring 1958.

5. Hemingway's final conversation with A.E. Hotchner. - A.E. Hotchner, *Papa Hemingway: A Personal Memoir* (New York: Random House, 1966; revised edition, Da Capo Press, 2004).

6. Muzafer Sherif and Carolyn Sherif - Robbers Cave experiment. - Muzafer Sherif et al., *The Robbers Cave Experiment: Intergroup Conflict and Cooperation* (Norman, OK: University of Oklahoma Press, 1961). The experiment took place in 1954 at Robbers Cave State Park, Oklahoma.

Chapter 7 - Autonomy

1. Vite, Patall, and Chen 2024 meta-analysis on autonomy and well-being. - Amanda Vite, Erika A. Patall, and Man Chen, "Relationships Between Experiences of Autonomy and Well(Ill)-Being for K-12 Youth: A Meta-Analysis," *Educational Psychology Review* 36, no. 4, Article 127 (2024). The meta-analysis examined 90 reports covering 101 studies.

2. Enslaved Black American craftsmen. - John Michael Vlach, *The Afro-American Tradition in Decorative Arts* (Cleveland: Cleveland Museum of Art, 1978). Philip D. Morgan, *Slave Counterpoint: Black Culture in the Eighteenth-Century Chesapeake and Lowcountry* (Chapel Hill: University of North Carolina Press, 1998).

3. Forced monasticism in medieval Europe. - C.H. Lawrence, *Medieval Monasticism: Forms of Religious Life in Western Europe in the Middle Ages* (London: Longman, 1984). Mayke de Jong, *In Samuel's Image: Child Oblation in the Early Medieval West* (Leiden: Brill, 1996).

4. Nietzsche - the will to power. - Friedrich Nietzsche, *Thus Spoke*

Zarathustra (1883-1885) and *Beyond Good and Evil* (1886).

Chapter 8 - Relatedness

1. Harvard Study of Adult Development. - Robert Waldinger and Marc Schulz, *The Good Life: Lessons from the World's Longest Scientific Study of Happiness* (New York: Simon & Schuster, 2023). The study began in 1938. Waldinger's quote from his 2015 TED Talk, "What Makes a Good Life? Lessons from the Longest Study on Happiness."

2. Sense of belonging predicts how meaningful life feels. - Nathaniel M. Lambert, Tyler F. Stillman, Joshua A. Hicks, Shanmukh Kamble, Roy F. Baumeister, and Frank D. Fincham, "To Belong Is to Matter: Sense of Belonging Enhances Meaning in Life," *Personality and Social Psychology Bulletin* 39, no. 11 (2013): 1418-1427.

3. Norwegian study of nearly 1,700 students - loneliness as strongest predictor of intention to leave. - Jan Arvid Haugan, Anne Frostad, and Per Egil Mjaavatn, "A Longitudinal Study of Factors Predicting Students' Intentions to Leave Upper Secondary School in Norway," *Social Psychology of Education* 22 (2019): 1489-1511.

4. Yale study - chocolate tasted better when shared. - Erica J. Boothby, Margaret S. Clark, and John A. Bargh, "Shared Experiences Are Amplified," *Psychological Science* 25, no. 12 (2014): 2209-2216.

5. Marie and Pierre Curie biographical details. - Susan Quinn, *Marie Curie: A Life* (New York: Simon & Schuster, 1995). Pierre's letter to Marie sourced to Marie Curie, *Pierre Curie*, pp. 72-77. Marie's diary entries and laboratory quote from Ève Curie, *Madame Curie: A Biography* (Garden City, NY: Doubleday, 1937), p. 249.

6. Edmund Hillary and Tenzing Norgay - Everest summit, May 29, 1953. - Edmund Hillary, *High Adventure* (London: Hodder and Stoughton, 1955). Tenzing Norgay with James Ramsey Ullman, *Tiger of the Snows*

(New York: G.P. Putnam's Sons, 1955). Also: Edmund Hillary, *View from the Summit* (2000).

7. Kipling Williams on ostracism. - Kipling D. Williams, *Ostracism: The Power of Silence* (New York: Guilford Press, 2001). Also: Kipling D. Williams, "Ostracism," *Annual Review of Psychology* 58 (2007): 425-452.

8. MIT Media Lab and OpenAI study (2025) - ChatGPT interactions and loneliness. - Cathy Mengying Fang et al., "How AI and Human Behaviors Shape Psychosocial Effects of Extended Chatbot Use: A Longitudinal Controlled Study," arXiv preprint arXiv:2503.17473, posted March 21, 2025. Related companion paper: Jason Phang et al., "Investigating Affective Use and Emotional Well-being on ChatGPT," arXiv:2504.03888, posted April 4, 2025.

9. Robert Weiss - six social provisions (1974). - Robert S. Weiss, "The Provisions of Social Relationships," in *Doing Unto Others*, ed. Zick Rubin (Englewood Cliffs, NJ: Prentice-Hall, 1974), 17-26.

10. Bay of Pigs invasion and Irving Janis on groupthink. - Irving L. Janis, *Victims of Groupthink: A Psychological Study of Foreign-Policy Decisions and Fiascoes* (Boston: Houghton Mifflin, 1972). Kennedy's quote attributed in Theodore Sorensen, *Kennedy* (New York: Harper & Row, 1965).

Chapter 9 - Transcendence

1. "If we have our own why of life, we shall get along with almost any how." - Friedrich Nietzsche, *Twilight of the Idols* (1889), "Maxims and Arrows," §12.

2. Viktor Frankl - Auschwitz observations, existential vacuum. - Viktor E. Frankl, *Man's Search for Meaning* (Boston: Beacon Press, 1959).

3. Erik Erikson on generativity. - Erik H. Erikson, *Childhood and Society*

(New York: W.W. Norton, 1950; revised 1963).

4. Harvard study on generativity and cognitive functioning. - George E. Vaillant, *Aging Well: Surprising Guideposts to a Happier Life from the Landmark Harvard Study of Adult Development* (Boston: Little, Brown, 2002).

5. Leo Tolstoy - suicidal depression, *Confession* quotes. - Leo Tolstoy, *A Confession* (1882).

6. Simone Weil biographical details. - Simone Pétrement, *Simone Weil: A Life* (New York: Pantheon Books, 1976). Richard Rees, *Simone Weil: A Sketch for a Portrait* (London: Oxford University Press, 1966).

Chapter 10 - The Recipe

1. John Kennedy Toole - *A Confederacy of Dunces*. - Pulitzer Prize for Fiction awarded posthumously in 1981. Published by Louisiana State University Press, 1980. His mother, Thelma Toole, persuaded Walker Percy to read the manuscript.

2. Sylvia Plath biographical details. - Heather Clark, *Red Comet: The Short Life and Blazing Art of Sylvia Plath* (New York: Knopf, 2020).

3. Julia Child biographical details. - Julia Child with Alex Prud'homme, *My Life in France* (New York: Knopf, 2006). Letters sourced from *As Always, Julia: The Letters of Julia Child and Avis DeVoto*, ed. Joan Reardon (Boston: Houghton Mifflin Harcourt, 2010).

4. Howard Hughes biographical details (revisited). - See endnote 26.

Chapter 13 - The New Status Games

1. Magnus Carlsen. - Standard biographical reference.

2. Linus Torvalds and Linux. - Linus Torvalds and David Diamond, *Just for Fun: The Story of an Accidental Revolutionary* (New York: HarperBusiness, 2001).

3. DHH (David Heinemeier Hansson) and Ruby on Rails; Evan You and Vue.js. - Standard technology history references.

4. Harriet Tubman - thirteen trips to the South. - Kate Clifford Larson, *Bound for the Promised Land: Harriet Tubman, Portrait of an American Hero* (New York: Ballantine Books, 2004).

Chapter 14 - Politics, Religion, and the Shape of Phase Three

1. Psychedelic research at Johns Hopkins, NYU, and Imperial College London. - Roland R. Griffiths et al., *Journal of Psychopharmacology* 30, no. 12 (2016): 1181-1197 (Johns Hopkins). Stephen Ross et al., *Journal of Psychopharmacology* 30, no. 12 (2016): 1165-1180 (NYU). Robin L. Carhart-Harris et al., *New England Journal of Medicine* 384 (2021): 1402-1411 (Imperial College London).

Chapter 15 - The Mission Starts Now

1. Youngstown Sheet and Tube / Campbell Works closure (1977). - Sean Safford, *Why the Garden Club Couldn't Save Youngstown* (Cambridge, MA: Harvard University Press, 2009). Staughton Lynd, *The Fight Against Shutdowns* (San Pedro, CA: Singlejack Books, 1982).

2. Soviet Union collapse - male life expectancy dropped by more than six years. - David A. Leon et al., "Huge variation in Russian mortality rates 1984-94," *The Lancet* 350, no. 9075 (1997): 383-388. Elizabeth Brainerd and David M. Cutler, "Autopsy on an Empire," *Journal of Economic Perspectives* 19, no. 1 (2005): 107-130.

3. Declining labor force participation among prime-age men. - Bureau of Labor Statistics, U.S. Department of Labor. Alan B. Krueger, "Where Have All the Workers Gone?," *Brookings Papers on Economic Activity* (Fall 2017).

Chapter 16 - What Can You Do For Society?

1. FDA approved OxyContin in 1995; more than 500,000 Americans

dead from opioids. - U.S. Food and Drug Administration records (OxyContin approval: December 1995). National Institute on Drug Abuse (NIDA), "Overdose Death Rates."

2. Bhutan - Gross National Happiness since the 1970s. - Introduced by King Jigme Singye Wangchuck in 1972. Centre for Bhutan Studies & GNH, "A Short Guide to Gross National Happiness Index" (Thimphu, 2012).

3. New Zealand Wellbeing Budget (2019). - New Zealand Treasury, "The Wellbeing Budget," May 30, 2019.

Chapter 17 - What Can You Do For Yourself?

1. Institute of Economic Affairs (UK) study - retirement and depression. - Gabriel H. Sahlgren, "Work Longer, Live Healthier," IEA Discussion Paper No. 46 (London: Institute of Economic Affairs, 2013).

2. Typesetting industry collapse and PageMaker (1985). - Aldus Page-Maker was released in July 1985. The typesetting industry declined rapidly through the late 1980s and 1990s.

Chapter 18 - Phase 4

1. Neuralink. - Founded by Elon Musk in 2016.

2. CRISPR. - Jennifer A. Doudna and Emmanuelle Charpentier, "The new frontier of genome engineering with CRISPR-Cas9," *Science* 346, no. 6213 (2014). Nobel Prize in Chemistry, 2020.

Chapter 19 - The Incomplete Opinions of Great Thinkers

1. Sam Altman's UBI study. - OpenResearch Basic Income study, personally funded by Sam Altman. Findings released in 2024. The study made $1,000 monthly payments to approximately 1,000 individuals in a total study of 3,000 people.

2. Dario Amodei, "Machines of Loving Grace." - Dario Amodei, "Ma-

chines of Loving Grace: How AI Could Transform the World for the Better," published on the Anthropic website, October 2024.

3. Elon Musk on Universal High Income and AI utopia. - Discussed across multiple platforms including the "Joe Rogan Experience" podcast, Episode #2054, November 2023, and posts on X (formerly Twitter) throughout 2024-2025.

4. Max Tegmark, *Life 3.0.* - Max Tegmark, *Life 3.0: Being Human in the Age of Artificial Intelligence* (New York: Knopf, 2017).

5. Yuval Noah Harari, *Homo Deus.* - Yuval Noah Harari, *Homo Deus: A Brief History of Tomorrow* (London: Harvill Secker, 2015; English edition, 2017).

Chapter 20 - The Unanswered Questions

1. North Korea nuclear weapons despite six-party talks. - The Six-Party Talks ran intermittently from 2003 to 2009, involving the U.S., China, Russia, Japan, South Korea, and North Korea. North Korea conducted its first nuclear test in October 2006.

Closing Thoughts

1. Naval Ravikant: "The reason to win the game is so that you can be free of it." - Naval Ravikant, *The Almanack of Naval Ravikant: A Guide to Wealth and Happiness*, ed. Eric Jorgenson (Magrudy's, 2020).

Index

C

D

E

F

About the Author

Joey Seeman dropped out of high school at fifteen to pursue a career in AI, built and sold his first AI company at sixteen, and became the youngest winner in all twenty-one seasons of *The Blox*, a business competition TV series. Today, he is the founder of two more AI companies and has personally implemented artificial intelligence for hundreds of businesses across dozens of industries.

He has spoken to audiences of thousands about AI implementation and has spent years helping thousands of business owners understand what AI is actually going to do to their companies, their employees, and their lives. Few people are closer to the truth of what happens when work disappears.

He has been in the rooms where people get replaced so he knows that what's coming next can either be the most extraordinary chapter in human history or the most empty one. He wrote this book to make sure we take the extraordinary path.

To learn more and connect with Joey Seeman, visit www.JoeySeeman.com